Spacecraft
for Astronomy

JOSEPH A. ANGELO, JR.

Facts On File
An imprint of Infobase Publishing

SPACECRAFT FOR ASTRONOMY

Facts On File, Inc.
An imprint of Infobase Publishing
132 West 31st Street
New York NY 10001

Library of Congress Cataloging in Publication Data
Angelo, Joseph A.
  Spacecraft for astronomy / Joseph A. Angelo, Jr.
    p. cm.
  Includes bibliographical references and index.
  ISBN 0-8160-5774-5
1. Space probes   Juvenile literature. 2. Astronomical instruments   Juvenile literature. 3. Astronomical observatories   Juvenile literature. 4. Space telescopes   Juvenile literature. I. Title
RL795.3.A54 2006
522'.2919   dc22   2006004875

Facts On File books are available at special discounts when purchased in bulk quantities for businesses, associations, institutions, or sales promotions. Please call our Special Sales Department in New York at (212) 967-8800 or (800) 322-8755.

You can find Facts On File on the World Wide Web at
http://www.factsonfile.com

Text design by Erika K. Arroyo
Cover design by Salvatore Luongo
Illustrations by Sholto Ainslie

Printed in the United States of America

VB FOF 10 9 8 7 6 5 4 3 2 1

This book is printed on acid-free paper.

In memory of my beloved daughter,
Jennifer April Angelo (April 26, 1975 to June 14, 1993)—
a beautiful young woman, whose promise-filled life was cut short by the
careless actions of others. Jenny, your radiant smile and dazzling emerald green
eyes provided all who knew you precious glimpses into the loving soul
of a very special person touched by the goodness and power of God.

Contents

✧ 12 Wrinkles in the Cosmic Microwave Background

185

✧ 13 Conclusion

208

Preface

*It is difficult to say what is impossible, for the dream of
yesterday is the hope of today and the reality of tomorrow.*

—Robert Hutchings Goddard

Frontiers in Space is a comprehensive multivolume set that explores
the scientific principles, technical applications, and impacts of space
technology on modern society. Space technology is a multidisciplinary
endeavor, which involves the launch vehicles that harness the principles
of rocket propulsion and provide access to outer space, the spacecraft that
operate in space or on a variety of interesting new worlds, and many different types of payloads (including human crews) that perform various
functions and objectives in support of a wide variety of missions. This
set presents the people, events, discoveries, collaborations, and important experiments that made the rocket the enabling technology of the
space age. The set also describes how rocket propulsion systems support
a variety of fascinating space exploration and application missions—
missions that have changed and continue to change the trajectory of
human civilization.

The story of space technology is interwoven with the history of astronomy and humankind's interest in flight and space travel. Many ancient
peoples developed enduring myths about the curious lights in the night
sky. The ancient Greek legend of Icarus and Daedalus, for example, por-
trays the age-old human desire to fly and to be free from the gravitational
bonds of Earth. Since the dawn of civilization, early peoples, including the
Babylonians, Mayans, Chinese, and Egyptians, have studied the sky and
recorded the motions of the Sun, the Moon, the observable planets, and
the so-called fixed stars. Transient celestial phenomena, such as a passing
comet, a solar eclipse, or a supernova explosion, would often cause a great
deal of social commotion—if not out right panic and fear—because these
events were unpredictable, unexplainable, and appeared threatening.

It was the ancient Greeks and their geocentric (Earth-centered) cosmology that had the largest impact on early astronomy and the emergence of Western Civilization. Beginning in about the fourth century B.C.E., Greek philosophers, mathematicians, and astronomers articulated a geocentric model of the universe that placed Earth at its center with everything else revolving about it. This model of cosmology, polished and refined in about 150 C.E. by Ptolemy (the last of the great early Greek astronomers), shaped and molded Western thinking for hundreds of years until displaced in the 16th century by Nicholaus Copernicus and a heliocentric (Sun-centered) model of the solar system. In the early 17th century, Galileo Galilei and Johannes Kepler used astronomical observations to validate heliocentric cosmology and, in the process, laid the foundations of the Scientific Revolution. Later that century, the incomparable Sir Isaac Newton completed this revolution when he codified the fundamental principles that explained how objects moved in the "mechanical" universe in his great work *The Principia.*

The continued growth of science over the 18th and 19th centuries set the stage for the arrival of space technology in the middle of the 20th century. As discussed in this multivolume set, the advent of space technology dramatically altered the course of human history. On the one hand, modern military rockets with their nuclear warheads redefined the nature of strategic warfare. For the first time in history, the human race developed a weapon system with which it could actually commit suicide. On the other hand, modern rockets and space technology allowed scientists to send smart robot exploring machines to all the major planets in the solar system (save for tiny Pluto), making those previously distant and unknown worlds almost as familiar as the surface of the Moon. Space technology also supported the greatest technical accomplishment of the human race, the Apollo Project lunar landing missions. Early in the 20th century, the Russian space travel visionary Konstantin E. Tsiolkovsky boldly predicted that humankind would not remain tied to Earth forever. When astronauts Neil Armstrong and Edwin (Buzz) Aldrin stepped on the Moon's surface on July 20, 1969, they left human footprints on another world. After millions of years of patient evolution, intelligent life was able to migrate from one world to another. Was this the first time such an event has happened in the history of the 14-billion-year-old universe? Or, as some exobiologists now suggest, perhaps the spread of intelligent life from one world to world is a rather common occurrence within the galaxy. At present, most scientists are simply not sure. But, space technology is now helping them search for life beyond Earth. Most exciting of all, space technology offers the universe as both a destination and a destiny to the human race.

Each volume within the Frontiers in Space set includes an index, a chronology of notable events, a glossary of significant terms and concepts, a helpful list of Internet resources, and an array of historical and current print sources for further research. Based upon the current principles and standards in teaching mathematics and science, the Frontiers in Space set is essential for young readers who require information on relevant topics in space technology, modern astronomy, and space exploration.

Acknowledgments

I wish to thank the public information specialists at the National Aeronautics and Space Administration (NASA), the National Oceanic and Atmospheric Administration (NOAA), the United States Air Force (USAF), the Department of Defense (DOD), the Department of Energy (DOE), the National Reconnaissance Office (NRO), the European Space Agency (ESA), and the Japanese Aerospace Exploration Agency (JAXA) who generously provided much of the technical materials used in the preparation of this series. Acknowledgement is made here for the efforts of Frank Darmstadt and other members of the editorial staff at Facts On File whose diligent attention to detail helped transform an interesting concept into a series of publishable works. The support of two other special people merits public recognition here. The first individual is my physician, Dr. Charles S. Stewart III, M.D., whose medical skills allowed me to successfully complete the series. The second individual is my wife, Joan, who, as she has for the past 40 years, provided the loving spiritual and emotional environment so essential in the successful completion of any undertaking in life, including the production of this series.

Introduction

Modern astrophysics has within its reach the ability to bring about one of the greatest scientific achievements ever—a unified understanding of the total evolutionary scheme of the universe. *Spacecraft for Astronomy* examines the important roles that astronomical instruments placed above Earth's atmosphere are playing in this remarkable intellectual revolution. In the middle of the 20th century, there was a fortuitous confluence of two streams of technical development: remote sensing and spaceflight. Through the science of remote sensing, scientists acquired sensitive instruments capable of detecting and analyzing radiation across the entire range of the electromagnetic spectrum. The arrival of spaceflight let astronomers and astrophysicists place sophisticated remote-sensing instruments above Earth's atmosphere on an extended mission basis. As shown in this book, the results of this technological union caused a major revolution in observational astronomy—similar to, if not more consequential, than when Galileo Galilei held his first primitive telescope up to the heavens in 1610. With his crude instrument, Galileo made a series of astounding discoveries, including mountains on the Moon, many new stars, and the four major moons orbiting Jupiter, which astronomers now refer to as the Galilean satellites in his honor. Galileo's work in observational astronomy provided direct evidence for the heliocentric Copernican hypothesis—the incendiary concept that ignited the scientific revolution of the 17th century.

Starting in the 1960s, an armada of increasingly more sophisticated orbiting astronomical observatories continued Galileo's legacy of discovery. Virtually all the information astronomers receive about distant celestial objects comes to them through observation of electromagnetic radiation. Cosmic-ray particles are an obvious exception, as are extraterrestrial material samples that are returned to Earth for scientific study. Similarly, sample return missions, like NASA's *Stardust,* also provide scientists the special opportunity of examining pristine pieces of extraterrestrial material firsthand. But for the most part, it is the revolution in remote-sensing technology that has provided incredibly new and important scientific insights about the universe, how it began, and where it is heading.

Each portion of the electromagnetic spectrum carries unique information about the physical conditions and processes in the universe. For example, infrared radiation reveals the presence of thermal emission from relatively cool objects and allows astronomers to peek through optically opaque clouds of dust and gas into the heart of stellar nurseries. Ultraviolet radiation provides astronomers information about the late stages of stellar evolution, while X-rays and gamma rays provide important information about extremely energetic celestial events and phenomena, such as supernovas, pulsars, and active galactic nuclei that may harbor supermassive black holes.

Spacecraft for Astronomy describes the historic events, scientific principles, and technical breakthroughs that allow complex orbiting astronomical observatories to revolutionize humans' understanding of the universe, its origin, and its destiny. In a manner reminiscent of Galileo's first use of the telescope to observe the night skies, optical astronomy has again enjoyed an enormous increase in benefits from large high-resolution optical systems operating above Earth's interfering atmosphere. For example, since its launch in 1990, the *Hubble Space Telescope* has made enormous contributions to astronomy. Similarly, X-ray astronomy has benefited greatly from the *Chandra X-ray Observatory,* gamma-ray astronomy from the *Compton Gamma Ray Observatory,* and infrared astronomy from the *Spitzer Space Telescope.* These four orbiting platforms, collectively referred to as NASA's Great Observatories, are just a few of the important spacecraft discussed in this book.

Spacecraft for Astronomy contains a specially selected collection of illustrations that shows historic, contemporary, and future space observatories. These illustrations allow readers to appreciate the tremendous technical progress that has occurred since the dawn of the Space Age in 1957 and what lies ahead. A generous number of sidebars are strategically positioned throughout the book to provide expanded discussions of fundamental scientific concepts and observational techniques. There are also capsule biographies of key physicists and astronomers to let the reader appreciate the human dimension in the development and operation of astronomical spacecraft.

It is especially important to recognize that throughout this century and beyond, sophisticated space-based astronomical instruments will continue to provide fundamental data that support many exciting scientific discoveries for the human race. Awareness of these technical activities should prove career inspiring to those students now in high school and college who will become the astronomers, astrophysicists, and cosmologists of tomorrow. Why are such career choices important? Future advances in astronomy, astrophysics, and cosmology represent a technical, social, and psychological imperative for the human race. Throughout history, people have pondered the origin, extent, and destiny of the universe. At this rather unique time in human history, spacecraft carrying astronomical instruments are making many important discoveries and providing vital new insights into the way the universe works. As a result, data from these spacecraft allow scientists to better address the fundamental questions that have occupied humans since their primitive beginnings. What is the nature of the universe? How did it begin, how is it evolving, and what will be its eventual fate?

Ever mindful of the impact of science and technology on society, *Spacecraft for Astronomy* examines the impact new astronomical discoveries have had on human development since the middle of the 20th century. This book then projects the expanded role space-based instruments will play throughout the remainder of this century and beyond. Who can now predict, for example, the incredible societal impact of discovering Earth-like worlds orbiting around nearby (within 100 or so light-years), Sun-like stars? Later in this century, as more advanced instruments allow scientists to examine candidate-habitable worlds, exobiologists might even be able to cautiously respond to the age-old philosophical question: Is Earth the unique life-bearing world in this vast universe?

This book also shows that the development of modern astronomical spacecraft did not occur without problems, issues, or major financial commitments. Selected sidebars within the book address some of the most pressing contemporary issues associated with the development and use of modern orbiting astronomical systems.

Spacecraft for Astronomy has been carefully crafted to help any student or teacher who has an interest in astronomy, astrophysics, or cosmology discover what space-based astronomical observatories and space exploration robots are, where they came from, how they work, and why they are so important. The back matter of this book contains a chronology, glossary, and an array of historical and current sources for further research. These should prove especially helpful for readers who need additional information on specific terms, topics, and events in astronomical observations and solar system investigations conducted by spacecraft operating above Earth's atmosphere.

From Petroglyphs to the *Spitzer Space Telescope*

"Man must rise above Earth—to the top of the atmosphere and beyond—for only thus will he fully understand the world in which he lives."

—Socrates

Astronomy is the branch of science that deals with celestial bodies and studies their size, composition, position, origin, and dynamic behavior. Dating back to antiquity, astronomy is one of humankind's oldest forms of science. Prior to the use of the telescope in the early 17th century, the practice of astronomy was generally limited to naked-eye observations, the use of now-defunct instruments such as the astrolabe and the cross staff, and the application of progressively more sophisticated forms of mathematics. The arrival of the telescope gave birth to optical astronomy—an influential discipline that greatly stimulated the scientific revolution and accelerated the blossoming of modern science.

The arrival of the Space Age with the launch of *Sputnik 1* in 1957 soon allowed astronomers to observe the universe across all portions of the electromagnetic spectrum and to visit the previously unreachable, alien worlds in our solar system. Today modern astronomy has many branches and subdisciplines, including astrometry, gamma-ray astronomy, X-ray astronomy, ultraviolet astronomy, visual (optical) astronomy, infrared astronomy, microwave astronomy, radar astronomy, and radio astronomy. Closely related scientific fields include high-energy astrophysics, cosmic-ray physics, neutrino physics, solar physics, condensed matter physics, space science, planetary geology, exobiology, and cosmology.

✧ From Petroglyphs to *The Almagest*

Prehistoric cave paintings (some up to 30 millennia old) provide a lasting testament that early peoples engaged in stargazing and incorporated such astronomical observations in their cultures. Early peoples also carved astronomical symbols in stones (petroglyphs) now found at numerous ancient ceremonial locations and ruins. Many of the great monuments and ceremonial structures of ancient civilizations have alignments with astronomical significance. Some people believe one of the oldest astronomical observatories is Stonehenge. Modern studies of Stonehenge suggest that the site probably served as an ancient astronomical calendar around 2000 B.C.E.

The Egyptians and the Maya both used the alignment of structures to assist in astronomical observations and the construction of calendars. Modern astronomers have discovered that the Great Pyramid at Giza, Egypt, has a significant astronomical alignment, as do certain Mayan structures—such as those found at Uxmal in the Yucatán, Mexico. Mayan astronomers were particularly interested in times (called "zenial passages") when the Sun crossed over certain latitudes in Central America. The Maya were also greatly interested in the planet Venus and treated the planet with as much reverence as the Sun. The Maya had a good knowledge of astronomy and could calculate planetary movements and eclipses over millennia.

For many ancient peoples, the motion of the Moon, the Sun, and the planets and the appearance of certain constellations of stars served as natural calendars that helped regulate daily life. Since these celestial bodies were beyond physical reach or understanding, various mythologies emerged along with native astronomies. Within ancient cultures, the sky became the home of the gods, and the Moon and Sun were often deified.

While no anthropologist really knows what the earliest human beings thought about the sky, the culture of the Australian Aborigines—which has been passed down for more than 40,000 years through the use of legends, dances, and songs—gives us a glimpse of how these early people interpreted the Sun, Moon, and stars. The Aboriginal culture is the world's oldest and most long-lived, and the Aboriginal view of the cosmos involves a close interrelationship between people, nature, and sky. Fundamental to their ancient culture is the concept of "the Dreaming"—a distant past when the spirit ancestors created the world. Aboriginal legends, dances, and songs express how in the distant past the spirit ancestors created the natural world and entwined people into a close relationship with the sky and with nature. Within the Aboriginal culture, the Sun is regarded as a woman. She awakes in her camp in the east each day and lights a torch that she then carries across the sky. In contrast, the Moon is seen as male.

Because of the coincidental association of the lunar cycle with the female menstrual cycle, they linked the Moon with fertility and consequently gave it a great magical status. These ancient peoples also regarded a solar eclipse as the male Moon uniting with the female Sun.

For the ancient Egyptians, Ra (also called Re) was regarded as the all-powerful Sun god who created the world and sailed across the sky each day. As a sign of his power, an Egyptian pharaoh would use the title "son of Ra." Within Greek mythology, Apollo was the god of the Sun, and his twin sister Artemis (Diana in Roman mythology) the Moon goddess.

From prehistory, astronomical observations have played a major role in the evolution of human cultures. The field of archaeoastronomy helps astronomers link contemporary knowledge of the heavens with the way ancient peoples viewed the sky and interpreted the mysterious objects they saw. Sophisticated astronomical computer software and star projectors (as found in planetariums) allow modern investigators to re-create how the night sky would have appeared to ancient peoples.

One of the tools of ground-based astronomy, then and now, is the use of astronomical constellations to help sort out and organize all the interesting "fixed" lights that appear in the night sky. (The ancient Greek astronomers called the wandering lights πλανετες (planetes), from which comes the word *planet*). A constellation is an easily identifiable (with the naked eye) configuration of the brightest stars in a moderately small region of the night sky. Originally, there was not one single set of constellations recognized by all astronomers. Rather, early astronomers in different regions of the world often defined and named the particular collections of bright stars they observed in relatively small regions of the sky after specific heroes, events, and creatures from their ancient cultures and mythologies.

The astronomical heritage of the constellations used in contemporary astronomy began about 2500 B.C.E. with the ancient stargazers of Mesopotamia. These early peoples used the stars to tell stories, to honor heroes and ferocious creatures—such as Orion the Hunter and Ursa Major (The Great Bear)—and to remind each new human generation that the heavens were the abode of the gods. In these Mesopotamian societies, (naked-eye) astronomy, mythology, religion, and cultural values were closely intertwined.

Early Greek astronomers adopted the constellations they found in Mesopotamia, embellished them with their own myths and religious beliefs, and eventually created a set of 48 ancient constellations. Eudoxus of Cnidus (c. 400–347 B.C.E.) was one of the first Greeks to formally codify these ancient constellations. About a century and a half later, Hipparchus of Nicaea (c. 190–120 B.C.E.), who is generally regarded by science historians as the greatest of the ancient Greek astronomers, reinforced the pioneering work of Eudoxus. Hipparchus's observational legacy included a star catalog completed in about 129 B.C.E. that contained 850 fixed stars. He also divided naked-eye observations into six magnitudes, ranging from the faintest (or least visible to the naked eye) to the brightest observable stars in the night sky (called first magnitude stars). Finally, the Greek astronomer Ptolemy (c. 100 –170 C.E.) cast the 48 ancient constellations in their present form around 150 C.E. in his great compilation of astronomical knowledge, *Syntaxis*. The early Greek astronomers were also keenly aware that these 48 constellations did not account for all the stars in the night sky. They even used a special word, *amorphotoi* (meaning "unformed"), to describe the spaces in the night sky populated by dim stars between the prominent groups of stars comprising the ancient constellations.

Today astronomers officially recognize all but one of these ancient star patterns. The somewhat cumbersome constellation Argo Navis has now been broken up into four new constellations (as discussed shortly). By convention, when astronomers formally refer to a constellation such as Centaurus in English, they use the proper name "The Centaur." Similarly, when astronomers wish to describe celestial objects within a particular constellation, they use the genitive form of the constellation's Latin name—as, for example, Alpha Centauri (α Cen) to describe the brightest (binary) star in the constellation Centaurus.

Astronomers in the Roman Empire were quite content to accept and use Greek celestial figures and constellations. Their primary contribution was the use of Roman (Latin) names for many familiar celestial objects—a tradition and heritage still followed by astronomers today. As part of the Pax Romana, the use of the 48 ancient Greek constellations spread throughout the civilized (Western) world centered around the Mediterranean Sea.

As the Roman Empire collapsed and the Dark Ages spread throughout western Europe, knowledge of Greek astronomy survived and began to flourish in Arab lands. Arab astronomers discovered Ptolemy's great work, translated it, and then renamed it *The Almagest.* By so doing, they preserved the astronomical heritage of the 48 ancient Greek constellations. Arab astronomers also refined and embellished the knowledge base of ancient Greek naked-eye astronomy by providing new star names and more precise observations.

✦ Modern Constellations

The astronomical heritage of the ancient constellations returned to Europe, just as people there were awakening from the Dark Ages, experiencing the Renaissance, and paving the way for the Scientific Revolution. As part of the explosive interest in astronomy that occurred at the start of the Scientific

Revolution, the German astronomer Johann Bayer (1572–1625) published the important work *Uranometria* in 1603. This book was the first major star catalog for the entire celestial sphere. Bayer charted more than 2,000 stars visible to the naked eye and introduced the practice of assigning Greek letters (such as alpha α, beta β, and gamma γ) to the main stars in each constellation, usually in an approximate (descending) order of their brightness. Expanding the legacy of 48 constellations from the ancient Greeks, Bayer named 12 new Southern Hemisphere constellations: Apus, Chamaeleon, Dorado, Grus, Hydrus, Indus, Musca (originally called Apis [bee] by Bayer), Pavo, Phoenix, Triangulum Australe, Tucana, and Volans.

Using a newly invented astronomical telescope, the Polish-German astronomer Johannes Hevelius (1611–87) filled in some of the empty spaces (amorphotoi) in the Northern Hemisphere of the celestial sphere by identifying the following new constellations: Canes Venatici, Lacerta, Lynx, and Leo Minor. Then, in the 18th century, the French astronomer Abbé Nicolas-Louis de Lacaille (1713–62) described 14 new constellations that he found in the Southern Hemisphere and named some of them after scientific artifacts and instruments emerging during the period. His newly identified constellations included: Antlia, Caelum, Circinus, Fornax, Horologium, Mensa, Microscopium, Norma, Octans, Pictor, Pyxis, Reticulum, Sculptor, and Telescopium. Lacaille was a precise technophile, so many of the names he carefully selected for his newly identified constellations were long and detailed. For example, Antlia (The Air Pump) honors the device invented by the British scientist Robert Boyle (1627–91), and Fornax actually means "The Laboratory Furnace." By international agreement, modern astronomers generally use shortened versions of Lacaille's original names—such as simply "furnace" instead of "laboratory furnace" for Fornax. This was done for ease in technical communications; it does not represent an attempt to detract from Lacaille's important work.

Lacaille and other astronomers of his era also dismantled the cumbersome ancient constellation Argo Navis (the ship of Jason and the Argonauts) and carved up the stars in this large Southern Hemisphere constellation into four smaller, more manageable ones whose names retain the original nautical theme: Carina (The keel), Puppis (The stern), Pyxis (The nautical compass), and Vela (The sails). Today astronomers use the 88 constellations that were officially recognized by the International Astronomical Union in 1929.

✧ Astrometry and the *Hipparcos* Spacecraft

Astrometry is the branch of astronomy that is concerned with the very precise measurement of the motion and position of celestial objects.

Astronomers often divide astrometry into two major categories: global astrometry (addressing positions and motions over large areas of the sky) and small-field astrometry (dealing with relative positions and motions that are measured within the area observed by a telescope—that is, within the instantaneous field of view of the telescope.)

Scientists involved in the practice of astrometry are often concerned with proper motion, nutation, and precession—phenomena that can cause the positions of celestial bodies to change over time. Astronomers participate in photographic astrometry when they accurately determine the positions of planets, stars, or other celestial objects with respect to the positions of reference stars as seen on photographic plates or high-resolution digital images.

Measuring the trigonometric parallax for a star provides the only completely reliable way of accurately determining distances in the local universe. Ground-based astrometric measurements are limited by several factors, including fluctuations in the atmosphere, limited sky coverage per observing site, and instrument flexure. Space-based astrometry helps overcome these disadvantages and supports global astrometry with many inherent advantages versus pointed (small-field) observations from the ground. The *Hipparcos* satellite was the first space-based astrometry mission and demonstrated that milliarcsecond accuracy is achievable by means of a continuously scanning satellite that observes two directions simultaneously.

The European Space Agency (ESA) launched the *Hipparcos* spacecraft in August 1989. *Hipparcos* measured the positions, distances, motions, brightness, and colors of stars. The spacecraft pinpointed more than 100,000 stars, 200 times more accurately than ever before. Astrometry has served as the bedrock science for the study of the universe since ancient times. Therefore, this successful mission represented a major leap forward that impacted every branch of astronomy. The name *Hipparcos* is an acronym for *Hi*gh *P*recision *Pa*rallax *Co*llecting *S*atellite; the word also closely resembles the name of the greatest naked-eye astrometrist in ancient Greece, Hipparchus of Nicaea. The other great astrometrist honored by this mission is the famous 16th-century Danish naked-eye observer Tycho Brahe (1546–1601), whose precise measurements of planetary orbits (especially Mars) helped Johannes Kepler (1571–1630) develop the three empirical laws of planetary motion.

This satellite was a pioneering space experiment dedicated to the precise measurement of positions, parallaxes, and proper motions of the stars. The goal was to measure the five astrometric parameters of some 120,000 primary program stars (an effort called the Hipparcos experiment) to a precision of some two to four milliarcseconds over a planned spacecraft lifetime of about two-and-one-half years. In addition, *Hipparcos* measured the astrometric and two-color photometric properties of some 400,000 additional stars (an effort called the Tycho experiment) to a somewhat lower astrometric precision. Once these mission goals were achieved, ESA flight controllers terminated communications with *Hipparcos* in August 1993. In June 1997, ESA published the final products of this mission as the Hipparcos Catalog and the Tycho Catalog.

✧ Galileo Galilei and Telescopic Astronomy

As the first astronomer to use a telescope to view the heavens, the Italian scientist Galileo Galilei (1564–1642) conducted early astronomical observations that helped inflame the Scientific Revolution of the 17th century. In 1610 he announced some of his early telescopic findings in the publication *Starry Messenger,* including the discovery of the four major moons of Jupiter (now called the Galilean satellites). The fact that they behaved like

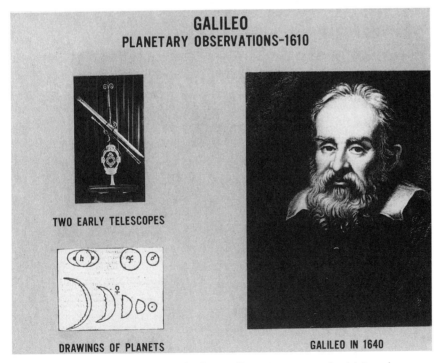

A 1640 portrait of Galileo Galilei—the fiery Italian astronomer, physicist, and mathematician who used his own version of the newly invented telescope to make detailed astronomical observations that helped inflame the Scientific Revolution in the 17th century. *(NASA)*

a miniature solar system stimulated his enthusiastic support for the helio-centric cosmology of Nicholas Copernicus (1473–1543). Unfortunately, this scientific work led to a direct clash with ecclesiastical authorities, who insisted on retaining the Ptolemaic system (with its geocentric cosmology) for a number of political and social reasons. By 1632 this conflict earned the fiery Galileo an Inquisition trial at which he was found guilty of heresy (for advocating the Copernican system) and confined to house arrest for the remainder of his life.

Galileo Galilei was born in Pisa on February 15, 1564. (Scientists and astronomers commonly refer to Galileo by only his first name). When he entered the University of Pisa in 1581, his father encouraged him to study medicine. His inquisitive mind soon became more interested in physics and mathematics than medicine. While still a medical student, he attended church services one Sunday. During the sermon, he noticed a chandelier swinging in the breeze and began to time its swing using his own pulse as a crude clock. When he returned home, he immediately set up an experiment that revealed the pendulum principle. After just two years of study, Galileo abandoned medicine and focused on mathematics and science. His change in career pathways also changed the entire trajectory of science.

In 1585 Galileo left the university without receiving a degree and focused his activities on the physics of solid bodies. The motion of fall-ing objects and projectiles intrigued him. Then, in 1589, he became a mathematics professor at the University of Pisa. Galileo was a brilliant lecturer and students came from all over Europe to attend his classrooms. This circumstance quickly angered many senior, but less capable, faculty members. To make matters worse, Galileo often used his tenacity, sharp wit, and biting sarcasm to win philosophical arguments at the university. His tenacious and argumentative personality earned him the nickname "The Wrangler."

In the late 16th century, European professors usually taught physics (then called natural philosophy) as an extension of Aristotelian philoso-phy and not as an observational, experimental science. Through his skill-ful use of mathematics and innovative experiments, Galileo changed that approach. His activities constantly challenged the 2,000-year tradition of ancient Greek learning. For example, Aristotle stated that heavy objects would fall faster then lighter objects. Galileo disagreed and held the oppo-site view that, except for air resistance, the two objects would fall at the same time regardless of their masses. Historians are not certain whether he personally performed the legendary musket ball–cannon ball drop experi-ment from the Leaning Tower in Pisa to prove this point. However, he did conduct a sufficient number of experiments with objects on inclined planes to upset Aristotelian physics and create the science of mechanics.

Throughout his life, Galileo was limited in his motion experiments by an inability to accurately measure small increments of time. Despite this impediment, he conducted many important experiments that produced remarkable insights into the physics of free fall and projectile motion. Less than a century later, Sir Isaac Newton (1642–1727) would build upon Galileo's work to create the universal law of gravitation and three laws of motion—the pillars of classical physics.

By 1592 Galileo's anti-Aristotelian research and abrasive behavior had offended his colleagues at the University of Pisa so much that they not so politely invited him to teach elsewhere. Later that year, Galileo moved to the University of Padua. This university had a more lenient policy of academic freedom, encouraged in part by the progressive government of the Republic of Venice. In Padua Galileo wrote a special treatise on mechanics to accompany his lectures. He also began teaching courses on geometry and astronomy. (At the time, the university's astronomy courses were primarily for medical students who needed to learn about what was known as medical astrology.)

In 1597 the German astronomer Johannes Kepler provided Galileo with a copy of Copernicus's book (even though the book was officially banned in Italy). Although Galileo had not previously been interested in astronomy, he discovered and immediately embraced the Copernican model. Galileo and Kepler continued to correspond until about 1610.

Between 1604 and 1605, Galileo performed his first public work involving astronomy. He observed the supernova of 1604 (in the constellation Ophiuchus) and used this astronomical event to refute the cherished Aristotelian belief that the heavens were immutable (unchangeable). He delivered this challenge on Aristotle's doctrine in a series of public lectures. Unfortunately, these well-attended lectures brought him into direct conflict with many of the university's pro-Aristotelian philosophy professors.

In 1609 Galileo learned that a new optical instrument (a magnifying tube) had just been invented in Holland. Within six months, Galileo devised his own version of the instrument. Then, in 1610, he turned this improved telescope to the heavens and started the age of telescopic astronomy. With his crude instrument, he made a series of astounding discoveries, including mountains on the Moon, many new stars, and the four major moons of Jupiter—now called the Galilean satellites in his honor. Galileo published these important discoveries in the book *Sidereus Nuncius* (Starry messenger). The book stimulated both enthusiasm and anger. Galileo used the moons of Jupiter to prove that not all heavenly bodies revolve around Earth. This provided direct observational evidence for the Copernican model—a cosmological model that Galileo now vigorously endorsed.

SUPERNOVAS

About once every 50 or so years, a massive star in the Milky Way Galaxy blows itself apart in a spectacular supernova explosion. Scientists regard supernovas as one of the most violent events in the universe. The force of a supernova explosion generates a blinding flash of radiation, as well as shock waves through interstellar space that are analogous to sonic booms. (Note that this is just a useful analogy, since sound waves cannot propagate through the vacuum of outer space.) Based on spectral classification activities in the 1930s, astrophysicists came to divide supernovas into two basic physical types: Type Ia and Type II. This classification is still used.

The Type Ia supernova involves the sudden explosion of a white dwarf star in a binary star system. A white dwarf is the evolutionary endpoint for stars with masses up to about five times the mass of the Sun. The remaining white dwarf has a mass of about 1.4 times the mass of the Sun and is about the size of Earth. A white dwarf in a binary star system will draw material off its stellar companion, if the two stars orbit close to each other. Mass capture hap-

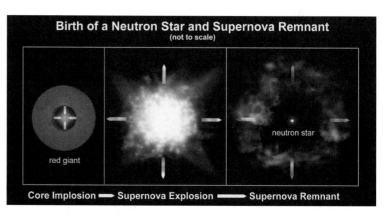

Birth of a Neutron Star and Supernova Remnant
(not to scale)

red giant

neutron star

Core Implosion ➤ Supernova Explosion ➤ Supernova Remnant

This is an artist's rendering that depicts the birth of a neutron star following a supernova explosion. (NASA/CXC)

pens because the white dwarf is a very dense object and exerts a very strong gravitational pull. Should the in-falling matter from a suitable companion star, such as a red giant, cause the white dwarf to exceed 1.4 solar masses (the Chandrasekhar limit), the white dwarf begins to experience gravitational collapse and creates internal conditions sufficiently energetic to support the thermonuclear fusion of elements like carbon. The carbon and other elements that make up the white dwarf begin to fuse uncontrollably, resulting in an enormous thermonuclear explosion that involves and consumes the entire star. Astronomers sometimes call this type of supernova a carbon-detonation supernova.

Unable to continue teaching old doctrine at the university, Galileo left Padua in 1610 and went to Florence. There he accepted an appointment as chief mathematician and philosopher to the grand duke of Tuscany, Cosimo II. He resided in Florence for the remainder of his life.

Because of *Sidereus Nuncius*, Galileo's fame spread throughout Italy and the rest of Europe. His telescopes were in demand, and he obligingly provided them to select European astronomers, including Kepler. In 1611

The Type II supernova is also called the massive star supernova. Stars that are five times or more massive than the Sun end their lives in a most spectacular way. The process starts when there is no longer enough fuel for the fusion process to occur in the core. When a star is functioning normally and in equilibrium, fusion takes place in its core, and the energy liberated by the thermonuclear reactions produces an outward pressure that combats the inward gravitational attraction of the star's great mass. But when a massive star that is going to become a supernova begins to die, it first swells into a red giant or supergiant—at least on the outside. In the interior, its core begins shrinking. As the star's core shrinks, the material becomes hotter and denser. Under these extremely heated and compressed conditions, a new series of thermonuclear reactions takes place, involving elements all the way up to, but not including, iron (Fe). Energy released from these fusion reactions temporarily halts the further collapse of the core.

However, this pause in the gravity-driven implosion of the core is only temporary. When the compressed core contains essentially only iron, all fusion ceases. (From nuclear physics, scientists know that iron nuclei are extremely stable and cannot fuse into higher atomic number elements.). At this point, the dying star begins the final phase of gravitational collapse. In less than a second, the core temperature rises to over 100 billion degrees Fahrenheit (55 billion degrees Celsius), as the iron nuclei are quite literally crushed together by the unrelenting influence of gravitational attraction. Then, this atom-crushing collapse abruptly halts due to the buildup of neutron pressure in the degenerate core material. The collapsing core recoils as it encounters this neutron pressure, and a rebound shock wave travels outward through the overlying material. The passage of this intense shock causes numerous nuclear reactions in the overlying outer material of the red giant, and the end result is a gigantic explosion. The shock wave also propels remnants of the original overlying material and any elements synthesized by various nuclear reactions out into space.

All but the central neutron star is blown away at speeds in excess of 31 million miles (50 million km) per hour. A thermonuclear shock wave races through the now-expanding stellar debris, fusing lighter elements into heavier ones and producing a brilliant visual outburst that appears as intense as the light of several billion Suns. Astronomers call the material that is exploded out into space a supernova remnant.

All that remains of the original star is a small, super-dense core composed almost entirely of neutrons—a neutron star. Because of the sequence of physical processes just mentioned, astrophysicists sometimes call a Type II supernova an implosion–explosion supernova. If the original star was very massive (perhaps 15 or more solar masses), even neutrons are not able to withstand the relentless collapse of the core's degenerate matter, and a black hole forms.

he proudly took one of his telescopes to Rome and let church officials personally observe some of these amazing celestial discoveries. While in Rome, he also became a member of the prestigious *Academia dei Lincei* (Lyncean academy). Founded in 1603, the academy was the world's first true scientific society.

In 1613 Galileo published his "Letters on Sunspots" through the academy. He used the existence and motion of sunspots to demonstrate that the Sun itself changes, again attacking Aristotle's doctrine of the immutability of the heavens. In so doing, he also openly endorsed the Copernican model. This started Galileo's long and bitter fight with ecclesiastical authorities. Above all, Galileo believed in the freedom of scientific inquiry. Late in 1615, Galileo went to Rome and publicly argued for the Copernican model. This public action angered Pope Paul V, who immediately formed a special commission to review the theory of Earth's motion.

Dutifully, the (unscientific) commission concluded that the Copernican theory was contrary to biblical teachings and possibly a form of heresy. Cardinal Robert Bellarmine (an honorable person who was later canonized) received the unenviable task of silencing the brilliant, but stubborn, Galileo. In late February 1616, he officially admonished Galileo to abandon his support of the Copernican hypothesis. Acting under direct orders from Pope Paul V, the cardinal made Galileo an offer he could not refuse. Galileo must never teach or write again about the Copernican model or else he would be tried for heresy and then imprisoned or possibly executed, like Giordano Bruno (1548–1600).

Apparently, Galileo got the message—at least so it seemed for a few years. In 1623 he published *Il Saggiatore* (The assayer). In this book, he discussed the principles for scientific research but carefully avoided support for Copernican theory. He even dedicated the book to his lifelong friend, the new pope, Urban VIII. However, in 1632 Galileo pushed his luck with the new pope to the limit by publishing *Dialogue on the Two Chief World Systems*. In this masterful (but satirical) work, Galileo had two people present scientific arguments to an intelligent third person, concerning the Ptolemaic and Copernican worldviews. The Copernican cleverly won these lengthy

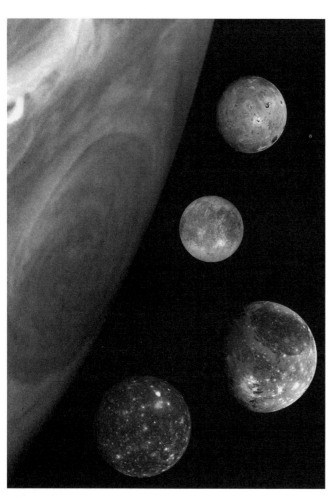

This is a scaled, composite image (scale factor: each pixel equals 9.3 miles [15 km]) of the major members of the Jovian system—collected by NASA's *Galileo* spacecraft during various flyby encounters in 1996 and 1997. Included in the interesting family portrait is the edge of Jupiter with its Great Red Spot, as well as Jupiter's four largest moons, called the Galilean satellites. From top to bottom, the moons shown are Io, Europa, Ganymede, and Callisto. *(NASA/JPL)*

arguments. Galileo represented the Ptolemaic system with an ineffective character that he called Simplicio.

For a variety of reasons, Pope Urban VIII regarded Simplicio as an insulting, personal caricature. Within months after the book's publication, the Inquisition summoned Galileo to Rome. Under threat of execution, the aging Italian scientist publicly retracted his support for the Copernican model on June 22, 1633. The Inquisition then sentenced him to life in prison, a term that he actually served under house arrest at his villa in Arceti (near Florence). Church authorities also banned the book *Dialogue,* but Galileo's supporters smuggled copies out of Italy, and the Copernican message continued to spread across Europe.

While under house arrest, Galileo worked on a less controversial area of physics. He published *Discourses and Mathematical Demonstrations Relating to Two New Sciences* in 1638. In this seminal work, he avoided astronomy and summarized the science of mechanics—including the very important topics of uniform acceleration, free fall, and projectile motion.

Through Galileo's pioneering work and personal sacrifice, the Scientific Revolution ultimately prevailed over misguided adherence to centuries of Aristotelian philosophy. Galileo never really opposed the church or its religious teachings. He did, however, come out strongly in favor of the freedom of scientific inquiry. Blindness struck the brilliant scientist in 1638. He died while imprisoned at home on January 8, 1642. Three-and-a-half centuries later, on October 31, 1992, Pope John Paul II formally retracted the sentence of heresy passed on him by the Inquisition.

✧ Balloons and Sounding Rockets Lead the Way into Space

Up until the Space Age, scientists were limited in their view of the universe by the Earth's atmosphere, which filters out most of the electromagnetic (EM) radiation from the rest of the cosmos. Even today ground-based astronomers are limited to just the visible portion of the EM spectrum and tiny portions of the infrared, radio, and ultraviolet regions. But space-based observatories allow them to examine the universe simultaneously in all portions of the EM spectrum. With spacecraft dedicated to astronomy and astrophysics, scientists have been able to study interesting cosmic phenomena and events in the visible, infrared, ultraviolet, X-ray, and gamma-ray portions of the EM spectrum. The rate of discovery has been nothing short of startling.

Prior to the launchings of the first artificial satellites in the late 1950s, scientists used high-altitude balloons and sounding rockets to take brief glimpses of the universe—tantalizing peeks that were unobstructed by most, if not all, Earth's intervening atmosphere. Two of the most significant discoveries were the existence of cosmic rays and the detection of extrasolar X-ray sources.

As a result of ionizing radiation measurements made during perilous high-altitude balloon flights between 1911 and 1913, the Austrian-American physicist Victor Francis Hess (1883–1964) discovered the existence of cosmic rays that continually bombard Earth from space. Detection of this very important astrophysical phenomenon earned him a share of the 1936 Nobel Prize in physics. Cosmic rays are very energetic nuclear particles that arrive at Earth from all over the Milky Way Galaxy. Scientists used Hess's discovery to turn Earth's atmosphere into a giant natural laboratory—a clever research approach that opened the door to many new discoveries in high-energy nuclear physics.

Hydrogen nuclei (protons) make up the highest percentage of the cosmic-ray population (approximately 85 percent), but these particles range over the entire periodic table of elements. Galactic cosmic rays are

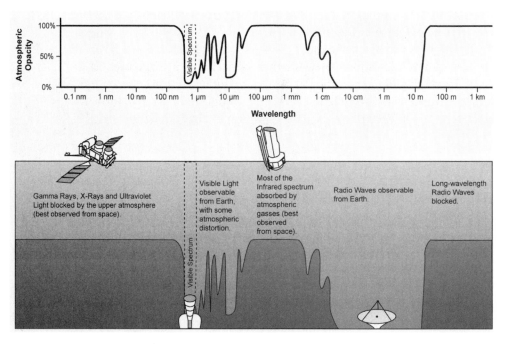

This graph shows how much of the electromagnetic spectrum (including visible light) gets through Earth's atmosphere at different wavelengths. For example, some near-infrared radiation wavelengths can reach high and dry mountaintop observatories. However, only space-borne infrared observatories provide an unimpeded view of the infrared universe. A similar situation occurs for incoming radiation from cosmic sources at gamma-ray, x-ray, or ultraviolet wavelengths, as well as long-wavelength radio waves. (*NASA, JPL, and Caltech*)

samples of material from outside the solar system and provide scientists direct evidence of phenomena that occur as a result of explosive processes in stars throughout the Milky Way Galaxy. Solar cosmic rays (mostly protons and alpha particles) are ejected from the Sun during solar flare events. Solar cosmic rays are generally lower in energy than galactic cosmic rays. As discussed in chapter 2, cosmic-ray astronomy is the branch of high-energy

ELECTROMAGNETIC SPECTRUM

When sunlight passes through a prism, it throws a rainbowlike array of colors onto a surface. This display of colors is called the visible spectrum. It represents an arrangement in order of wavelength of the narrow band of electromagnetic (EM) radiation to which the human eye is sensitive.

The electromagnetic spectrum comprises the entire range of wavelengths of electromagnetic radiation, from the shortest-wavelength gamma rays to the longest-wavelength radio waves. The entire EM spectrum includes much more than meets the eye.

The names applied to the various regions of the EM spectrum are (going from shortest to longest wavelength) gamma ray, X-ray, ultraviolet, visible, infrared, and radio. EM radiation travels at the speed of light (i.e., about 186,000 miles [300,000 km] per second) and is the basic mechanism for energy transfer through the vacuum of outer space.

One of the most interesting discoveries of 20th-century physics is the dual nature of electromagnetic radiation. Under some conditions electromagnetic radiation behaves like a wave, while under other conditions it behaves like a stream of particles, called photons. The tiny amount of energy carried by a photon is called a quantum of energy (plural: quanta). The word *quantum* comes to us from the Latin language and means "little bundle."

The shorter the wavelength, the more energy is carried by a particular form of EM radiation. All things in the universe emit, reflect, and absorb electromagnetic radiation in their own distinctive ways. The way an object does this provides scientists with special characteristics, or signatures, that can be detected by remote-sensing instruments. For example, the spectrogram shows bright lines for emission or reflection and dark lines for absorption at selected EM wavelengths. Analyses of the positions and line patterns found in a spectrogram can provide information about the object's composition, surface temperature, density, age, motion, and distance.

Emission spectroscopy involves analytical spectroscopic methods that use the characteristic electromagnetic radiation emitted when materials are subjected to thermal or electrical sources for purposes of identification. These thermal or electrical sources excite the molecules or atoms in the sample of material to energy levels above the ground state. As the molecules or atoms return to ground state from these higher energy states, electromagnetic radiation is emitted in discrete, characteristic wavelengths or emission lines. The pattern and intensity of these emission lines create a unique emission spectrum that enables the analyst to identify the substance.

astrophysics that uses cosmic rays to provide information on the origin of the chemical elements through nucleosynthesis during stellar explosions.

Collaborating with Professor Bruno Benedetto Rossi (1905–93) at MIT and other researchers in the early 1960s, Riccardo Giaconni (1931–   ) designed novel X-ray detection instruments for use on sounding rockets and, eventually, spacecraft. In June 1962, one of Giacconi's sounding-rocket experiments proved especially significant. His team placed an instrument package on a sounding rocket in an attempt to perform a brief search for possible solar-induced X-ray emissions from the lunar surface. To the team's great surprise, the instruments accidentally detected the first-known X-ray source outside the solar system, a star called Scorpius X-1. Previous rocket-borne experiments by personnel from the U.S. Naval Research Laboratory that started in 1949 had detected X-rays from the Sun, but Scorpius X-1, the so-called X-ray star, was the first source of X-rays known to exist beyond the solar system. It is the brightest and most persistent nontransient X-ray source in the sky. This fortuitous discovery is often regarded as the beginning of X-ray astronomy—an important field within high-energy astrophysics. Giacconi shared the 2002 Nobel Prize in physics for his pioneering contributions to astrophysics.

The following analogy illustrates the significance of Giacconi's efforts in X-ray astronomy and also serves as a graphic testament to the rapid rate of progress in astronomy brought on by the Space Age. By historic coincidence, Giacconi's first, relatively crude X-ray telescope was approximately the same length and diameter as the first astronomical (optical) telescope used by Galileo Galilei in 1610. Over a period of about four centuries, optical telescopes improved in sensitivity by 100 million times, as their technology matured from Galileo's first telescope to the capability of NASA's *Hubble Space Telescope.* About 40 years after Giacconi tested the first X-ray telescope on a sounding rocket, NASA's magnificent *Chandra X-ray Observatory* provided scientists a leap in measurement sensitivity of about 100 million. Much as Galileo's optical telescope revolutionized observational astronomy in the 17th century, Giacconi's X-ray telescope triggered a modern revolution in high-energy astrophysics in the 20th century. (Chapter 6 discusses X-ray astronomy.)

To complement space-based astronomical observations at infrared (IR) wavelengths, NASA has also used airborne observatories—that is, infrared telescopes carried by airplanes that reach altitudes above most of the infrared radiation absorbing water vapor in Earth's atmosphere. The first such observatory was a modified Lear Jet, the second a modified C-141A Starlifter jet transport. NASA called the C-141A airborne observatory the Kuiper Airborne Observatory (KAO) in honor of the Dutch-American astronomer Gerard Peter Kuiper (1905–73). The KAO's three-foot (0.9-m-) aperture reflector was a Cassegrain telescope designed primarily for observations in the one to 500 micrometer wavelength spectral range. This aircraft collected infrared images of the universe from 1974 until its retirement in 1995.

✧ Lyman Spitzer Jr. and the Vision of Space-Based Astronomy

In 1946, more than a decade before the launch of the first artificial satellite, the American astrophysicist Lyman Spitzer Jr. (1914–97) proposed the development of a large, space-based observatory that could operate unhindered by distortions in Earth's atmosphere. His vision ultimately became NASA's *Hubble Space Telescope,* which was launched in 1990. Spitzer was a renowned astrophysicist who made major contributions in the areas of stellar dynamics, plasma physics, and thermonuclear fusion, as well as space-based astronomy. NASA launched the *Space Infrared Telescope Facility* in 2003 and renamed this sophisticated new space-based infrared telescope the *Spitzer Space Telescope* in his honor.

Spitzer was born on June 26, 1914, in Toledo, Ohio. He attended Yale University, where he earned his bachelor's degree in physics in 1935. Following a year at Cambridge University, he entered Princeton University, where he earned his master's degree in 1937 and then a doctorate in astrophysics in 1938. His mentor and doctoral adviser was the famous American astronomer Henry Norris Russell (1877–1957). After getting his Ph.D., Spitzer spent a year as a postdoctoral fellow at Harvard University, after which he joined the faculty of Yale University (1939).

During World War II, Spitzer performed underwater sound research, working with a team that led the development of sonar. When the war was over, he returned for a brief time to teach at Yale University. In 1946, more than a decade before the first artificial satellite (*Sputnik 1*) was launched into space and 12 years before NASA was created, Spitzer proposed the pioneering concept of placing an astronomical observatory in space—where it could observe the universe over a wide range of wavelengths and not have to deal with the blurring (or absorbing) effects of Earth's atmosphere. He further proposed that a space-based telescope would be able to collect much clearer images of very distant objects in comparison to any

The American astronomer Lyman Spitzer Jr. was a strong advocate for space-based astronomy. *(NASA/Denise Applewhite/Princeton University)*

ground-based telescope. To support these views, Spitzer wrote "Astronomical Advantages of an Extra-Terrestrial Observatory." In this visionary paper, he enumerated the advantages of putting a telescope in space. He would then invest a considerable amount of his time over the next five decades to serve as an enthusiastic lobbyist for a telescope in space, both with members of the U.S. Congress and with fellow scientists. His efforts were instrumental in the development of the *Hubble Space Telescope.*

In 1947 Spitzer received an appointment as chairman of Princeton University's astrophysical sciences department. Accepting this appointment, he succeeded his doctoral adviser, Henry Norris Russell. Spitzer also became the director of Princeton's observatory. While at Princeton, he made many contributions to the field of astrophysics. For example, he thoroughly investigated interstellar dust grains and magnetic fields, as well as the motions of star clusters and their evolution. He also studied regions of star formation and was among the first astrophysicists to suggest that bright stars in spiral galaxies formed recently from the gas and dust there. Finally, he accurately predicted the existence of a hot galactic halo surrounding the Milky Way Galaxy.

In 1951 Spitzer founded the Princeton Plasma Physics Laboratory (originally called Project Matterhorn by the U.S. Atomic Energy Commission). This laboratory became Princeton University's pioneering program

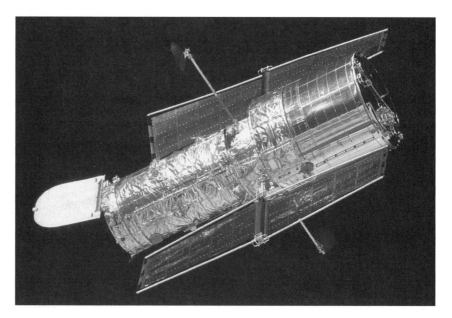

The *Hubble Space Telescope* with its aperture door open, floating majestically in space. This image was taken by the crew of the space shuttle *Discovery* in February 1997, during the STS–82 servicing mission. *(NASA/JSC)*

in controlled thermonuclear research. Spitzer promoted efforts to harness nuclear fusion as a clean source of energy and remained the laboratory's director until 1967. Then, in 1952, Spitzer became the Charles A. Young Professor of Astronomy at Princeton. He retained that prestigious title for the rest of his life.

From 1960 to 1962, Spitzer served as president of the American Astronomical Society. As the fledging U.S. space program emerged in the 1960s, Spitzer's visionary idea for space-based astronomy finally began to look more promising. In 1962 he led a program to design an observatory that would orbit the Earth and study the ultraviolet light from space. Earth's atmosphere normally blocks ultraviolet (UV) light, so scientists cannot study UV emissions from cosmic sources using ground-based facilities. This proposed observatory eventually became NASA's *Copernicus* spacecraft, which operated successfully between 1972 and 1981. (Chapter 8 discusses ultraviolet astronomy.)

In 1965 the National Academy of Sciences established a committee to define the scientific objectives for a proposed large space telescope, and the academy selected Spitzer to chair this committee. At the time, many astronomers did not support the idea of a large space-based telescope. They were concerned, for example, that the cost of an orbiting astronomical facility would reduce the government's financial support for ground-based astronomy. Spitzer invested a great personal effort to convince members of the scientific community, as well of the U.S. Congress, that placing a large telescope into space had great scientific value. In 1968 the first step in making Spitzer's vision of putting a large telescope in space came true. That year NASA launched its highly successful Orbiting Astronomical Observatory series of scientific spacecraft. An observatory spacecraft is a space robot that does not travel to a destination to explore. Instead this type of robot spacecraft travels in an orbit around Earth or around the Sun, from where the observatory can view distant celestial targets unhindered by the blurring and obscuring effects of Earth's atmosphere.

Through the early 1970s, Spitzer continued to lobby NASA and the U.S. Congress for the development of a large space telescope. Finally, in 1975, NASA, along with the European Space Agency, began development of what would eventually become the *Hubble Space Telescope*. In 1977, due in large part to Spitzer's unflagging efforts, the U.S. Congress approved funding for the construction of NASA's *Space Telescope*—an orbiting facility eventually named the *Hubble Space Telescope* in honor of the great American astronomer Edwin P. Hubble (1889–1953). In 1990, more than 50 years after Spitzer first proposed placing a large telescope into space, NASA used the space shuttle to successfully deploy the *Hubble Space Telescope* in orbit around Earth. Refurbished several times on orbit-servicing

NASA's Orbiting Astronomical Observatory

NASA launched a series of large astronomical observatories in the late 1960s to significantly broaden scientific understanding of the universe. The first successful large observatory placed in Earth's orbit was the *Orbiting Astronomical Observatory 2* (*OAO-2*), nicknamed *Stargazer*, which was launched on December 7, 1968. In its first 30 days of operation, *OAO-2* collected more than 20 times the celestial ultraviolet data than had been acquired in the previous 15 years of sounding rocket launches. *Stargazer* also observed Nova Serpentis for 60 days after its outburst in 1970. These observations confirmed that mass loss by the nova was consistent with theory. NASA's *Orbiting Astronomical Observatory 3*, named *Copernicus* in honor of the famous Polish astronomer Nicholas Copernicus, was launched successfully on August 21, 1972. This satellite provided much new data on stellar temperatures, chemical compositions, and other properties. It also gathered data on the black hole candidate Cygnus X-1, so named because it was the first X-ray source discovered in the constellation Cygnus.

missions by the space shuttle, the *Hubble Space Telescope* still provides scientists with stunning images of the universe and still produces amazing new discoveries. (The *Hubble Space Telescope* is discussed in chapter 4.)

At the age of 82, Spitzer passed away on March 31, 1997, in Princeton, New Jersey. NASA launched a new space telescope on August 25, 2003. This space-based observatory consists of a large and lightweight telescope and three cryogenically cooled science instruments capable of studying the universe at near- to far-infrared wavelengths. Incorporating state-of-the-art infrared detector arrays, and launched into an innovative Earth-trailing solar orbit, the observatory is orders of magnitude more capable than any previous space-based infrared telescope. NASA named this new facility the *Spitzer Space Telescope* in honor of Spitzer's vision of and contributions to space-based astronomy. (The *Spitzer Space Telescope* is discussed in chapter 7.)

✧ Robot Spacecraft in Service to Astronomy

Each portion of the electromagnetic spectrum (that is, radio waves, infrared radiation, visible light, ultraviolet radiation, X-rays, and gamma rays) brings astronomers and astrophysicists unique information about the universe and the objects within it. For example, certain radio frequency signals help scientists characterize cold molecular clouds. The cosmic microwave background represents the fossil radiation from the big bang, the enormous ancient explosion considered by most scientists to have started the present universe about 14 billion years ago. The infrared portion of the spectrum provides signals that let astronomers observe non-visible objects such as near-stars (brown dwarfs) and relatively cool stars. Infrared radiation also helps scientists peek inside dust-shrouded stellar nurseries (where new stars are forming) and unveil optically opaque regions at the core of the Milky Way Galaxy. Ultraviolet radiation provides astrophysicists special information about very hot stars and quasars, while visible light helps observational astronomers characterize planets, main sequence stars, nebulae, and galaxies. Finally, the collection of X-rays and gamma rays by space-based observatories brings scientists unique information about high-energy phenomena, such as supernovas, neutron stars, and black holes—whose presence is inferred by intensely energetic radiation emitted from extremely hot material as it swirls in an accretion disk before crossing the particular black hole's event horizon.

Scientists recognized that they could greatly improve their understanding of the universe if they could observe all portions of the electromagnetic spectrum. As the technology for space-based astronomy matured toward the end of the 20th century, NASA created the Great Observatories Program. This important program involved a series of four highly sophisticated space-based astronomical observatories—each carefully designed with state-of-the-art equipment to gather "light" from a particular portion (or portions) of the electromagnetic spectrum. An observatory spacecraft is a robot spacecraft that does not have to travel to a celestial destination to explore it. Instead, the observatory spacecraft occupies a special orbit around Earth or an orbit around the Sun, from where it can

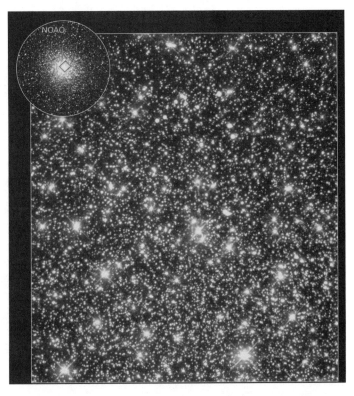

Looking into the heart of a glittering swarm of stars, this *Hubble Space Telescope* image (taken in November 2002) unveils the central region of the globular cluster M22, a 12- to 14-billion-year-old grouping of stars in the constellation Sagittarius. The inset photograph (upper left, taken in June 1995) shows the entire globular cluster of about 10 million stars. *(NASA/GSFC for HST image; Nigel A. Sharp, REU program/ AURA/NOAO/NSF for inset photo)*

observe distant targets free of the obscuring and blurring effects of Earth's atmosphere. Infrared observatories should also operate in orbits that minimize interference from large background thermal radiation sources like Earth and the Sun.

NASA initially assigned each Great Observatory a development name and then renamed the orbiting astronomical facility to honor a famous scientist. The first Great Observatory was the *Space Telescope*, which became the *Hubble Space Telescope* (*HST*). It was launched by the space shuttle in 1990 and then refurbished on-orbit through a series of subsequent shuttle

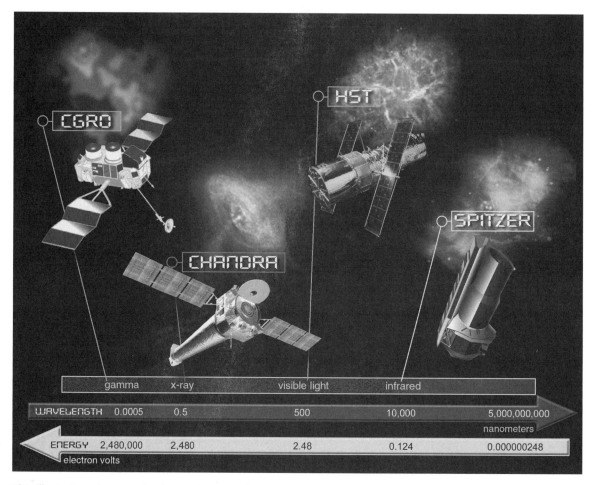

This illustration shows each of NASA's Great Observatories and the region of the electromagnetic spectrum from which the particular space-based astronomy facility collects scientific data. From left to right (in order of decreasing photon energy and increasing wavelength): the *Compton Gamma Ray Observatory*; the *Chandra X-ray Observatory*; the *Hubble Space Telescope*; and the *Space Infrared Telescope Facility*, now called the *Spitzer Space Telescope*. (NASA)

missions. With constantly upgraded instruments and improved optics, this long-term space-based observatory is designed to gather light in the visible, ultraviolet, and near-infrared portions of the spectrum. This spacecraft honors the American astronomer Edwin Powell Hubble (1889–1953). NASA is now examining plans for another (possibly robotic) refurbishment mission that would keep the *HST* operating for several more years until being replaced by the *James Webb Space Telescope* around 2011.

The second Great Observatory was the *Gamma Ray Observatory,* which NASA renamed the *Compton Gamma Ray Observatory* (*CGRO*) following its launch by the space shuttle in 1991. Designed to observe high-energy gamma rays, this observatory

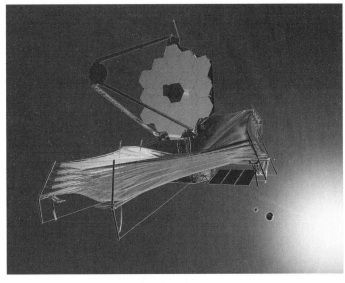

This is an artist's rendering of NASA's *James Webb Space Telescope,* which is scheduled for launch about 2011. *(NASA/ Northrop Grumman)*

collected valuable scientific information from 1991 to 1999 about some of the most violent processes in the universe. NASA renamed the observatory to honor the American physicist and Nobel laureate Arthur Holly Compton (1892–1962). The *CGRO*'s scientific mission officially ended in 1999. The following year, NASA mission managers commanded the massive spacecraft to perform a controlled de-orbit burn. This operation resulted in a safe reentry in June 2000 and the harmless impact of surviving pieces in a remote portion of the Pacific Ocean.

NASA originally called the third observatory in this series the *Advanced X-ray Astrophysics Facility.* It renamed this observatory the *Chandra X-ray Observatory* (*CXO*) to honor the Indian-American astrophysicist and Nobel laureate Subrahmanyan Chandrasekhar (1910–95). The observatory spacecraft was placed into a highly elliptical orbit around Earth in 1999. The *CXO* examines X-ray emissions from a variety of energetic cosmic phenomena, including supernovas and the accretion disks around suspected black holes, and should operate until at least 2009.

The fourth and final member of NASA's Great Observatory Program is the *Space Infrared Telescope Facility.* NASA launched this observatory in 2003 and renamed it the *Spitzer Space Telescope* (*SST*) to honor the American astrophysicist Lyman Spitzer Jr. (1914–97). The sophisticated infrared observatory provides scientists a fresh vantage point from which to study processes that have until now remained mostly in the dark, such as the formation of galaxies, stars, and planets. The *SST* also serves as an important technical bridge to NASA's Origins Program—an ongoing attempt to scientifically address such fundamental questions as "Where did we come from?" and "Are we alone?"

High-Energy Astrophysics: Meeting the Universe Face-to-Face

A stronomy addresses fundamental questions that have occupied humans since their primitive beginnings. What is the nature of the universe? How did it begin, how is it evolving, and what will be its eventual fate? As important as these questions are, there is another motive for astronomical studies. Since the 17th century, when Sir Isaac Newton's studies of celestial mechanics helped him to formulate the three basic laws of motion and the universal law of gravitation, the sciences of astronomy and physics have been intertwined.

Scientists call the study of the nature and physics of stars and star systems astrophysics. This discipline provides the theoretical framework for understanding astronomical observations. At times scientists use astrophysics to predict celestial phenomena, such as black holes, before these phenomena have even been observed by astronomers. The laboratory of outer space makes it possible to investigate large-scale physical processes that cannot be duplicated in a terrestrial laboratory. Although the immediate, tangible benefits to humankind from progress in astrophysics cannot easily be measured or predicted, the opportunity to extend scientific understanding of the workings of the universe is really an integral part of the development of modern civilization.

Starting in the 20th century, astrophysicists formed a special intellectual bond with nuclear physicists. When astrophysicists study the universe on a grand scale and use space-based instruments to look back in time to just after universe was created in a gigantic ancient explosion (called the big bang), they uncover interesting insights into the nature of matter, energy, and the fundamental forces that also govern nature on the smallest, subatomic scales. When nuclear physicists use extremely energetic particle accelerators to explore the inner secrets of matter and energy on a subatomic scale, they often shed light on some amazing new

X-ray data from NASA's *Chandra X-Ray Observatory*, ESA's *XMM-Newton* spacecraft and the German ROSAT have provided direct evidence about the catastrophic destruction of a star that wandered too close to a supermassive black hole. This artist's rendering depicts how this event might have occurred. A close encounter with another star put the doomed star on a path that took it near a supermassive black hole. The enormous gravity of the giant black hole then stretched the star until it was torn apart. Because of the momentum and energy of the accretion process, only a few percent of the disrupted star's mass was actually swallowed by the black hole, while the rest of the star's mass was flung away into the surrounding galaxy (called RX J1242-11). *(NASA/CXC/M. Weiss)*

aspect of of the fundamental forces that govern the behavior of exotic celestial objects, such as quasars, neutron stars, quark stars, and black holes. Similarly, while an astrophysicist ponders the mysterious physics that occur at the center of a black hole, a nuclear physicist is busy trying to unravel the secrets of how matter and energy behave as quarks interact to form protons and neutrons.

✧ Elementary Particles—It Is a Truly Small, Small World

An elementary particle is a fundamental constituent of matter. In the early portion of the 20th century, the atomic nucleus model of Ernest Rutherford (1871–1937) and James Chadwick's (1891–1974) discovery of the neutron suggested a universe consisting of three elementary particles: the proton, neutron, and electron. That simple model is still very useful in describing nuclear phenomena because many of the other elementary particles are very short-lived and appear only briefly during nuclear reactions.

The British scientist Sir J. J. Thomson (1856–1940) discovered the existence of the electron in the late 1890s. An electron (symbol: e) is a stable elementary particle with a unit negative electrical charge (1.602×10^{-19} coulomb) and a rest mass $(m_e)_0$ of 1/1,837 that of a proton, or about 20×10^{-31} pounds (9.109×10^{-31} kg). Electrons surround the positively charged nucleus and determine the chemical properties of the atom. Positively charged electrons, or positrons, also exist.

The proton (symbol: p) is a stable elementary nuclear particle with a single positive charge and a rest mass of about 3.7×10^{-27} pounds (1.672×10^{-27} kg), which is about 1,837 times the mass of an electron. A single proton makes up the nucleus of an ordinary (or) light hydrogen atom. Protons are also constituents of all other nuclei. The atomic number (Z) of an atom is equal to the number of protons in its nucleus.

The neutron (symbol: n) is an uncharged elementary particle with a mass slightly greater than that of the proton. Chadwick discovered this interesting elementary particle in 1932. The neutron is found in the

An artist's rendering that compares the relative sizes of a neutron star, a quark star, and the Grand Canyon in Northern Arizona. The Grand Canyon is a giant gorge formed by the Colorado River. The scenic canyon extends westward 217 miles (349 km) from the mouth of the Little Colorado River to Lake Mead, with a width of four to 18 miles (6.5–30 km) and a maximum depth of one mile (1.6 km). *(NASA/CXC)*

nucleus of every atom heavier than ordinary hydrogen. A free neutron is unstable, with a half-life of about 10 minutes, and decays into an electron, a proton, and a neutrino. Neutrons play an important role in astronomy and astrophysics. For example, a neutron star is a very small (typically 12.5 to 18.5 miles [20 to 30 km] in diameter), super-dense stellar object—the gravitationally collapsed core of a massive star that has undergone a supernova explosion. Astrophysicists believe that pulsars are rapidly spinning young neutron stars that possess intense magnetic fields.

To explain the strong forces that exist inside the nucleus between the nucleons, physicists developed quantum chromodynamics (QCD) in the 1960s and introduced the quark as the basic building block of hadrons—the class of heavy subatomic particles (including neutrons and protons) that experiences this strong interaction or short-range strong nuclear force. Physicists call the other contemporary family of elementary particles with finite masses leptons—light particles (including electrons) that participate in electromagnetic and weak interactions but not the strong nuclear force.

According to quantum theory, a photon is the elementary bundle or packet of electromagnetic radiation, such as a photon of light, an X-ray, or a gamma ray. As presently postulated by physicists, photons have no mass and travel at the speed of light. The energy (E) of the photon is equal to the product of the frequency (ν) of the electromagnetic radiation and Planck's constant (h). Planck's constant is a fundamental constant in nature and has a value of 6.626×10^{-34} joule-second. Physicists express the frequency in hertz.

Quantum theory is one of the two pillars of modern physics. (The other pillar of modern physics is Albert Einstein's theory of relativity.) Quantum theory was started in 1900 by the German physicist Max Karl Planck (1858–1947) when he proposed that all electromagnetic radiation is emitted and absorbed in quanta, or discrete energy packets. According to quantum theory, the energy of a photon is related to its frequency by the equation: $E = h\nu$, where E is the photon energy, ν is the frequency, and h is Planck's constant. In 1905 Albert Einstein (1879–1955) used quantum theory to explain the photoelectric effect. Einstein's Nobel Prize–winning work assumed that light propagated in quanta or photons. In 1913 the Danish physicist Niels Bohr (1885–1962) combined quantum theory with Ernest Rutherford's nuclear atom hypothesis. Bohr's work resolved the difficulties with the Rutherford atomic model and also served as the major intellectual catalyst that promoted the emergence of quantum mechanics in the 1920s.

Quantum mechanics is the physical theory that emerged in the 1920s and 1930s from Max Planck's original quantum theory. Within the realm of quantum mechanics, the Heisenberg uncertainty principle and the Pauli exclusion principle provided a framework that dictated how particles behaved at the atomic and subatomic levels. The Heisenberg uncertainty principle is an important physical principle which states that it is impossible to simultaneously know the momentum and precise position of a subatomic particle. This basic principle of quantum mechanics also implies that the physicist cannot precisely know both energy and time on a microscopic scale. In 1927 the German physicist Werner Heisenberg (1901–76) suggested that the uncertainties of position (Δx) and momentum (Δp) of a subatomic particle are related by the following inequality: $\Delta x\, \Delta p \geq h/4\pi$, where h is Planck's constant.

The Pauli exclusion principle is another basic principle in quantum mechanics. This principle states that no two fermions (particles such as quarks, protons, neutrons, and electrons) may exist in the same quantum state. The physical principle was initially proposed in the mid-1920s by the German physicist Wolfgang Pauli (1900–58). The Pauli exclusion principle implies that no two electrons in a given atom can have the same quantum number values for energy, spin, and angular momentum. As a result of Heisenberg's and Pauli's work, quantum mechanics spawned three important variants: wave mechanics, matrix mechanics, and relativistic quantum mechanics.

A quark is any of six extremely tiny elementary particles (diameter less than 3×10^{-18} feet [1×10^{-18} m]). Its existence was independently postulated in 1963 by physicists Murray Gell-Mann (1929–   ) and George Zweig (1937–   ) as being the basic building block of matter residing inside hadrons, such as neutrons and protons. There are six types or "flavors" of quark: the up (u), down (d), strange (s), charmed (c), top (t), and bottom (b). Physicists postulate that all quarks have spin ½. Another interesting feature of quarks is that they have fractional electrical charges. For example, the up quark carries an electrical charge of $+ \frac{2}{3}$e, while the down quark carries one of $- \frac{1}{3}$e. The symbol e represents the basic charge of an electron. Quarks possess another property called color, for which physicists assign three possibilities: blue, green, and red. Corresponding to each quark, there is an antiquark with oppositely signed electric charge (namely, $+\frac{1}{3}$e or $-\frac{2}{3}$e) and anti-blue, anti-green, and anti-red colors. Please note that physicists selected the name of this characteristic and the colors—blue, green, and red—rather whimsically.

The quark quality of color is important in QCD, but it has nothing to do with the visible portion of the electromagnetic spectrum. As presently assumed in the standard model of nuclear physics, hadrons (baryons and mesons) are made of doublet and triplet combinations of quarks. The neutron, for example, consists of one up and two down quarks, while a proton consists of one down and two up quarks. Physicists think exchanging gluons (massless particles serving as the carriers of the strong nuclear force) holds quarks together. Today some astrophysicists suggest that deep within neutron stars are states of condensed matter best characterized as a soup of free quarks.

A quark star is postulated as a very highly condensed (degenerate) matter star consisting of free quarks. Astrophysicists speculate that a quark star has a density that lies between the density of a neutron star and that of a black hole. When the quark star was first suggested in the 1980s, few astrophysicists anticipated really finding one in nature. However, in April 2002, NASA's *Chandra X-ray Observatory* detected an exotic object known as RX J185635-375. With a diameter of about six miles (10 km), this object is too small to be a neutron star. So, some scientists suggest that the object is a candidate quark star—a very highly condensed matter star whose nucleons (that is, protons and neutrons) have become so squeezed together by the self-attractive force of gravity that the nucleons themselves burst open, releasing their constituent quarks.

A quantum chromodynamics is a quantum field theory (gauge theory) that describes the strong force interactions among quarks and antiquarks, as these minute, subnuclear-size elementary particles exchange gluons. The strong force is one of the four fundamental forces in nature (discussed in the next section), and physicists also refer to it as the nuclear force or the strong nuclear force. It is the force that binds quarks together to form baryons (3 quarks) and mesons (2 quarks). For example, QCD addresses how the nucleons of everyday matter, neutrons and protons, consist of the quark combinations *uud* and *udd*, respectively. Here the symbol *u* represents a single up quark, while the symbol *d* represents a single down quark. The force that holds nucleons together to form an atomic nucleus is envisioned as a residual interaction between the quarks inside each nucleon.

Quantum electrodynamics (QED) is the branch of modern physics independently copioneered by Richard Feynman (1918–88), Julian Schwinger (1918–94), and Sinitiro Tomonaga (1906–79). QED combines quantum mechanics with the classical electromagnetic theory of the 19th-century Scottish physicist James Clerk Maxwell (1831–79). QED also provides a quantum mechanical description of how electromagnetic radiation (in the form of photons) interacts with charged matter.

Quantum gravity is a hypothetical theory of gravity that would result from the successful merger of general relativity (gravity) and quantum mechanics. Although no such theory yet exists, there is a need for this theory to help physicists address problems involving the very early universe—that is, from the big bang up to Planck time (approximately 10^{-43} second)—and the behavior of black holes. The graviton is the hypothetical particle in quantum gravity that plays a role similar to that of the photon in quantum electrodynamics.

Although cosmology is discussed in chapter 12, it is beneficial to introduce the notion of quantum cosmology here as part of the discussion on quantum mechanics. *Quantum cosmology* is a collective term used by scientists to describe the various theories found within modern cosmology that address the physics of the very early universe, using quantum gravity—that is, the combination of quantum mechanics and gravity. Ultimately, quantum cosmology requires a viable theory of quantum gravity. In an interesting application of experimental and theoretical research from modern high-energy physics, particle physicists have suggested that the early universe experienced a specific sequence of phenomena and phases following the big bang explosion. Physicists generally refer to this sequence of events as the standard cosmological model. Right after the big bang, the present universe reached an incredibly high temperature—physicists offer the unimaginable value of about 10^{32} K. During this period, called the quantum gravity epoch, the force of gravity, the strong nuclear force, and a composite electroweak force all behaved as a single unified force.

At Planck time (about 10^{-43} second after the big bang), the force of gravity assumed its own identity. Scientists call the ensuing period the grand unified epoch. While the force of gravity functioned independently during this period, the strong nuclear force and the composite electroweak force remained together and continued to act like a single force. Today physicists apply various grand unified theories in their efforts to model and explain how the strong nuclear force and the electroweak force functioned as one during this period in the early universe.

Then, about 10^{-35} second after the big bang, the strong nuclear force separated from the electroweak force. By this time, the expanding universe had "cooled" to about 10^{28} K. Physicists call the period between about 10^{-35} s and 10^{-10} s the electroweak epoch. During this epoch, the weak nuclear force and the electromagnetic force became separate entities, as the composite electroweak force disappeared. From this time forward, the universe contained the four basic forces known to physicists today—namely, the force of gravity, the strong nuclear force, the weak nuclear force, and the electromagnetic force.

Following the big bang and lasting up to about 10^{-35} s, there was no distinction between quarks and leptons. All the minute particles of matter were similar. However, during the electroweak epoch, quarks and leptons became distinguishable. This transition allowed quarks and antiquarks to eventually become hadrons, such as neutrons and protons, as well as their antiparticles. At 10^{-4} s after the big bang, in the radiation-dominated era, the temperature of the universe cooled to 10^{12} K. By this time, most of the hadrons had disappeared because of matter-antimatter annihilations. The surviving protons and neutrons represented only a small fraction of the total number of particles in the universe—the majority of which were leptons, such as electrons, positrons, and neutrinos. However, like most of the hadrons before them, the majority of the leptons also soon disappeared as a result of matter-antimatter interactions.

At the beginning of the matter-dominated era, about 200 seconds (or some three minutes) following the big bang, the expanding universe cooled to a temperature of 10^9 K, and small nuclei, such as helium, began to form. Later, when the expanding universe reached an age of about 500,000 years (or some 10^{13} seconds old), the temperature dropped to 3,000 K, allowing the formation of hydrogen and helium atoms. Interstellar space is still filled with the remnants of this primordial hydrogen and helium.

Eventually, density inhomogeneities allowed the attractive force of gravity to form great clouds of hydrogen and helium. Because these clouds also experienced local density inhomogeneities, gravitational attraction formed stars and then slowly gathered groups of these stars into galaxies. As gravitational attraction condensed the primordial (big bang) hydrogen and helium into stars, nuclear reactions at the cores of the more massive stars created heavier nuclei up to and including iron. Supernova explosions then occurred at the end of the lives of many of these early, massive stars. These spectacular explosions produced all atomic nuclei with masses beyond iron and then scattered the elements into space. The expelled stardust would later combine with interstellar gas and eventually create new stars, along with their planets. For example, all things animate and inanimate here are on Earth are the natural by-products of these ancient astrophysical processes.

✧ Fundamental Forces in Nature

At present physicists recognize the existence of four fundamental forces in nature: gravity, electromagnetism, the strong nuclear force, and the weak nuclear force. Gravity and electromagnetism are part of everyday experience. These two forces have an infinite range, which means they exert their influence over great distances. The other two forces are much less familiar because both only act within the realm of the atomic nucleus. But out of sight does not mean insignificant or out of mind. Rather, the strong and weak nuclear forces govern the mass-energy transformations that light up the stars and galaxies.

Gravity is the attractive force that tugs on people and holds them on the surface of Earth. Gravity also keeps the planets in orbit around the Sun and causes the formation of planets, stars, and galaxies. Sir Isaac Newton, the world's first astrophysicist, introduced his theory of gravity

Sir Isaac Newton—The World's First Astrophysicist

Sir Isaac Newton (1642–1727) was the brilliant though introverted British astrophysicist and mathematician whose law of gravitation, three laws of motion, development of calculus, and design of a new type of reflecting telescope make him one of the greatest scientific minds in human history. Through the patient encouragement and financial support of the British mathematician Sir Edmund Halley (1656–1742), Newton published his great work, the *Philosophiae Naturalis Principia Mathematica* (Mathematical principles of natural philosophy; also known as *Principia Mathematica* or *The Principia*), in 1687. This monumental book transformed the practice of physical science and completed the scientific revolution started by Nicholas Copernicus, Johannes Kepler, and Galileo Galilei.

Newton was born prematurely in Woolsthorpe, Lincolnshire, on December 25, 1642 (Julian calendar) His father died before Newton's birth, a tragedy that contributed to his very unhappy childhood. Throughout his life, Newton would not tolerate criticism, remained hopelessly absent-minded, and often tottered on the verge of emotional collapse. Historians claim that Newton laughed only once or twice in his entire life. Yet, despite the gravity of his personality, Newton is still considered by many to be the greatest human intellect who ever lived. His brilliant work in physics, astronomy, and mathematics combined the discoveries of Copernicus, Kepler, and Galileo. Newton's universal law of gravitation and his three laws of motion fulfilled the Scientific Revolution and dominated science for the next two centuries.

In 1653 Newton's mother returned to the farm at Woolsthorpe and removed her son from school so he could practice farming. Fortunately for science, Newton failed miserably as a farmer. So, in June 1661, he departed Woolsthorpe for Cambridge University. Newton graduated without any particular honors from Cambridge with a bachelor's degree in 1665. Following graduation he returned to his mother's farm in Woolsthorpe to avoid the plague, which had broken out in London. For the next two years of self-imposed exile, Newton pondered mathematics and physics.

By Newton's own account, one day on the farm he saw an apple fall to the ground and began to wonder if the same force that pulled on the apple also kept the Moon in its place. As part of the Scientific Revolution, heliocentric cosmology was becoming widely accepted (except where banned on political or religious grounds). But the mechanism for planetary motion around the Sun remained elusive and unexplained.

By 1667 the plague epidemic had subsided, and Newton returned to Cambridge, receiving his master of arts degree the following year. Around this time, he also constructed the first working reflecting telescope, a device that now carries his name. The Newtonian telescope uses a parabolic mirror to collect light. The primary mirror then reflects the collected light by means of an internal secondary mirror to an external focal point at the side of the telescope's tube. This new telescope design earned Newton a great deal of professional acclaim, including eventual membership in the Royal Society.

(continues)

(continued)

In 1669 Newton became the Lucasian Professor of Mathematics at Cambridge. Shortly after his election to the Royal Society (in 1671), Newton published his first paper in that society's transactions. While an undergraduate, Newton had used a prism to refract a beam of white light into its primary colors (red, orange, yellow, green, blue, and violet.) Newton reported this important discovery to the Royal Society. However, Newton's pioneering work was immediately attacked by Robert Hooke (1635–1703), an influential member of the society.

This was the first in a series of bitter disputes between Hooke and Newton that ended only with Hooke's death in 1703. Newton only skirmished lightly then quietly retreated. This was Newton's continuing pattern of avoiding direct conflict. When he became famous later in his life, Newton would start a controversy, withdraw, and then secretly manipulate others who would then carry the brunt of the battle against Newton's adversary. Newton's famous conflict with the German mathematician Gottfried Leibniz (1646–1716) over the invention of calculus followed precisely such a pattern.

In August 1684, Edmund Halley convinced the reclusive genius to address the following puzzle about planetary motion: What type of curve does a planet describe in its orbit around the Sun, assuming an inverse square law of attraction? To Halley's delight, Newton immediately responded: "an ellipse." Halley pressed on and asked Newton how he knew the answer to this important question. Newton nonchalantly informed Halley that he had already done the calculations years ago (in 1666) but claimed to have misplaced them. So Newton promised to send Halley another set as soon as he could.

Newton sent Halley his *De Motu Corporum in Gyrum* (On the motion of bodies in an orbit; 1684), in which he demonstrated that the force of gravity between two bodies is directly proportional to the product of their masses and inversely proportional to the square of the distance between them—a principle now called Newton's universal law of gravitation. Halley was astounded and begged Newton to carefully document all of his work on gravitation and orbital mechanics. Through Halley's patient encouragement and financial support, Newton published *Principia Mathematica* in 1687. In this work, Newton gave the world his famous three laws of motion and the universal law of gravitation. This monumental work transformed physical science and completed the scientific revolution started by Copernicus, Kepler, and Galileo. Many historians consider the *Principia Mathematica* as the greatest scientific accomplishment of the human mind.

For all his brilliance, Newton was also extremely fragile. After completing the *Principia Mathematica*, he drifted away from physics and astronomy and eventually suffered a serious nervous disorder in 1693. Newton was so bitter about his quarrels with Hooke that he waited until 1704 (the year after Hooke's death) to publish his other major work, *Opticks*. Queen Anne knighted him in 1705. Even late in life, Newton could not tolerate controversy. But now, as president of the Royal Society, he skillfully maneuvered younger scientists to fight his intellectual battles. In this manner, the world's first astrophysicist continued to rule the scientific landscape until his death in London on March 20, 1727.

in 1687 when he published his universal law of gravitation in the *Principia Philosophiae Naturalis Mathematica* (Mathematical principles of natural philosophy; also known as *Principia Mathematica* or *The Principia*). Newton based his perception of gravity as an attractive force—one that always pulled but never pushed away—on the pioneering efforts of Galileo Galilei, who was the first scientist to carefully investigate the way objects move. Galileo paid special attention to the behavior of objects in free fall. He also meticulously studied the motion of projectiles fired out of a cannon and followed their ballistic trajectories under the influence of gravity.

For Newton, gravity was an inherent property of matter, directly proportional to an object's mass. In the solar system, the Sun reaches out across enormous distances and essentially pulls smaller masses, such as planets, asteroids, and comets, into orbit around it, using the (attractive) force of gravity. This simple model worked well within the framework of Newtonian mechanics. During the 18th and 19th centuries, physicists and astronomers remained comfortable with Newton's concept of gravity.

But Newton's gravity applecart was upset early in the 20th century when another brilliant physicist, Albert Einstein, introduced his theory of special relativity (1905). After presenting special relativity, Einstein soon discovered a serious problem with Newton's concept of gravity as an attractive force. One of the major postulates in special relativity is that the speed of light is the speed limit of all energy and matter in the universe. Energy, in the form of radiation or rapidly moving particles, cannot transmit (propagate) across the universe any faster than the speed of light, which is approximately 186,450 miles (300,000 km) per second in a vacuum. So, Einstein was puzzled, because Newton's theory of gravity assumed that the Sun's attractive force instantaneously transmitted to the planets, at a speed much faster than the speed of light. (Earth is about eight light-minutes away from the Sun.) Was gravity unique in its ability to reach across the universe, or did masses attract each other for a different reason? This intriguing question challenged Einstein for the next decade and led him to publish his theory of general relativity in 1916.

Einstein's general relativity replaced Newton's notion of space as a vast emptiness with nothing but the invisible force of gravity to rule the motion of matter with the concept of space-time, an ephemeral fabric that grips matter and directs its course through the universe. According to Einstein's theory, the space-time fabric spans the entire universe and is intimately connected to all matter and energy within it. Hypothetically, when a mass sits in this space-time fabric, it deforms the fabric itself, changing the shape of space and altering the passage of time around it.

In the case of the Sun, the space-time fabric would curve around it, creating a dip in space-time. As other celestial objects, such as the planets, comets, and asteroids, travel across the space-time fabric, they respond to this dip created by the Sun and follow the curve in space-time, traveling around and around the Sun. As long as they do not slow down, the planets would maintain regular orbits around the Sun, neither spiraling in toward it nor flying off into outer space. Much more massive, compact objects like neutron stars and black holes would create large dips or distortions in the space-time fabric or continuum.

Einstein's general relativity imagines that gravity is generated when matter distorts the space-time continuum. To better visualize this important concept, imagine a scientist placing a heavy iron ball in the center of a large, sturdy sheet of rubber that is stretched horizontally and securely fastened along its edges. Once the heavy ball is gently placed in the middle of the rubber sheet, the sheet will stretch a certain amount, forming a dip or depression. The depth of this dip in the sheet depends on the mass of the heavy ball and the stiffness of the rubber sheet. If the researcher now rolls some small balls across the rubber sheet at different points, he or she would observe them curve in toward the central mass. But in this thought experiment, the balls are not pulled in by (or attracted to) the heavy ball's gravity. Rather, the small balls simply follow the curve in the rubber sheet (that is, the dip in the "space-time continuum") caused by the presence of the heavy central mass. Einstein's concept of gravity as related to the curvature of space-time is explored further in chapter 12.

However, for day-to-day experiences, Newton's concept of gravity as an attractive force still remains quite useful. For example, the dreaded bathroom scale, which faithfully reports all manner of eating indiscretions and undesired "weight" gains, depends on the concept of Earth's gravity tugging on a person's mass, producing a force called that person's weight. Here Earth's mass is considered concentrated at the center of the planet, and the person is assumed located on Earth's surface. After a little manipulation with Newton's second law of motion, the weight becomes that person's mass times the local acceleration due to gravity. Standing on the Moon, that person's mass will not change, but his or her weight will be about one-sixth of its value on Earth. Why? The reason is because the Moon is smaller than Earth and has as a lower value for the acceleration due to gravity (about one-sixth that of Earth). So, astronomers and astrophysicists still use Newton's universal law of gravitation in many astronomical problems that lie within the realm of classical physics.

The second fundamental force is electromagnetism. This force is responsible for the way matter generates and responds to electricity and magnetism. Modern civilization is structured around systems engineered to use electricity and magnetism, including those for transportation, tele-communications, electric motors, lighting, computers, and data storage.

The other two forces, the strong nuclear force and the weak nuclear force, operate within the realm of the atomic nucleus and involve elementary particles. These forces are beyond our day-to-day experience and were essentially unknown to the physicists of the 19th century, who had a good classical understanding of both the universal law of gravity and the fundamental principles of electromagnetism.

The strong nuclear force operates at a range of about 3×10^{-15} feet $(1 \times 10^{-15}$ m) and holds the atomic nucleus together. The weak nuclear force has a range of about 3×10^{-17} feet $(1 \times 10^{-17}$ m) and is responsible for processes like beta decay that tear nuclei and elementary particles apart.

What is important to recognize here is that whenever anything happens in the universe—that is, whenever an object experiences a change in motion—the event takes place because one or more of these fundamental forces are involved.

✧ NASA's High-Energy Astronomy Observatory

NASA's High-Energy Astronomy Observatory (HEAO) program consisted of a series of three space-based astronomy missions designed to map and image X-rays. A survey of cosmic gamma-ray sources was also performed. Between 1977 and 1979, NASA placed three different HEAO spacecraft in approximately circular orbits around Earth. These large, massive spacecraft carried equally massive scientific payloads and provided pioneering support to high-energy astrophysics and astronomy.

NASA launched *HEAO-1* on August 12, 1977, from Cape Canaveral, using an Atlas-Centaur rocket. *HEAO-1* was primarily a survey mission, dedicated to systematically mapping the X-ray sky every six months. The six-sided *HEAO-1* spacecraft had a mass of 5,615 pounds (2,552 kg), was 18.6 feet (5.68 m) high, and 8.76 feet (2.67 m) in diameter.

NASA engineers and scientists designed the *HEAO-1* spacecraft to continue the X-ray and gamma-ray studies initiated by previous astronomical spacecraft, such as NASA's *Orbiting Astronomical Observatory 3*, which was also called the *Copernicus* spacecraft. The HEAO-1 mission was specifically designed to meet the following scientific goals: to map and survey the celestial sphere for X-ray and gamma-ray sources in the energy range of 150 eV to 10 MeV, to establish the size and precise location of X-ray sources in order to determine the contribution of discrete sources to the X-ray background, and to measure the time variations of X-ray source. The *HEAO-1* spacecraft had a mission lifetime from August 12, 1977, to

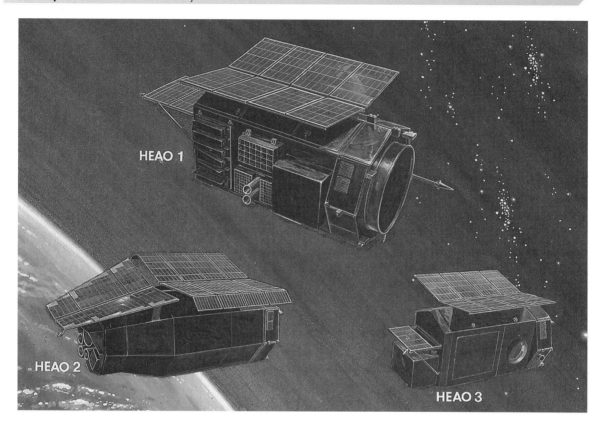

This artist's concept shows the family of NASA's High-Energy Astronomy Observatory (HEAO) spacecraft. *HEAO-1* was launched on August 12, 1977; *HEAO-2* on November 13, 1978; and *HEAO-3* on September 20, 1979. *(NASA/ Marshall Space Flight Center)*

January 9, 1979, and operated in a 268-mile- (432-km-) altitude circular orbit around Earth, with an inclination of 23 degrees and a period of 93.5 minutes.

NASA launched the *HEAO-2* spacecraft on November 13, 1978, from Cape Canaveral using an Atlas-Centaur rocket. Following a successful launch, NASA renamed *HEAO-2,* calling the spacecraft the *Einstein Observatory* to honor the great physicist Albert Einstein. The *Einstein Observatory* carried the first fully imaging X-ray telescope ever placed into orbit. The 6,890-pound (3,130-kg) spacecraft was shaped like a hexagonal prism, had a height of 18.6 feet (5.68 m), and a diameter of 8.76 feet (2.67 m).

The primary objectives of this mission were imaging and spectrographic studies of specific X-ray sources and studies of the diffuse X-ray background. The *HEAO 2* spacecraft was identical in basic design to the

HEAO 1 vehicle, with the addition of reaction wheels and associated electronics to enable the observatory to point its X-ray telescope at sources to an accuracy of within one arc minute.

The instrument payload had a mass of 3,190 pounds (1,450 kg). A large grazing-incidence X-ray telescope provided images of sources that were then analyzed by four interchangeable instruments mounted on a carousel arrangement that could be rotated into the focal plane of the telescope. The four instruments were a solid-state spectrometer (SSS), a focal plane crystal spectrometer, an imaging proportional counter, and a high-resolution imaging detector. The X-ray telescope had an angular resolution of a few arc seconds, a field-of-view of tens of arc minutes, and a sensitivity of approximately 1,000 times greater than any X-ray instrument used in previous space missions. Also included in the science payload were a monitor proportional counter, which viewed the sky along the telescope axis, a broadband filter, and objective grating spectrometers that could be used in conjunction with focal plane instruments.

The major scientific objectives of the *Einstein Observatory* were to locate accurately and to examine with high spectral resolution X-ray sources in the energy range 0.2 to 4.0 keV and to perform high-spectral-sensitivity measurements with both high- and low-dispersion spectrographs. The science payload also performed high-sensitivity measurements of transient X-ray behavior. Downlink telemetry was accomplished at a data rate of 6.5 kilobits per second (kb/s) for real-time data and 128 kb/s for either of two tape-recorder systems. An attitude control and determination subsystem were used to point and maneuver the spacecraft. The spacecraft also used gyroscopes, Sun sensors, and star trackers as sensing devices for pointing information and attitude determination. The *Einstein Observatory* operated from November 1978 to April 1981 in an approximately circular orbit around Earth at an altitude of 292 miles (470 km), with an inclination of 23.5 degrees and a period of 94 minutes. Using the innovative X-ray telescope designed by Nobel laureate Riccardo Giacconi and his associates, the *HEAO-2/Einstein*

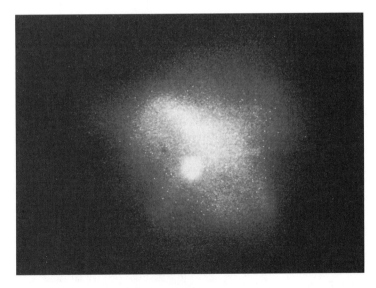

An X-ray image of the Crab Nebula taken by the *Einstein Observatory.* The image is dominated by a pulsar, which appears as a bright point due to its X-ray emissions. This image depicts a portion of the remnant of a supernova that was recorded by Chinese astronomers in 1054 C.E. *(NASA/Marshall Space Flight Center)*

spacecraft performed trailblazing high-resolution spectroscopy and morphological studies of supernova remnants, including the Crab Nebula.

HEAO-3 was the third of NASA's High-Energy Astronomy Observatories. NASA launched *HEAO-3* on September 20, 1979, from Cape Canaveral, using an Atlas-Centaur rocket. The spacecraft operated in an approximately circular orbit, with an average altitude of 308 miles (495 km), an inclination of 43.6 degrees, and a period of 94.5 minutes. Like *HEAO-1,* the *HEAO-3* scientific spacecraft was a survey mission, but this time it examined the hard X-ray and gamma-ray portion of the electromagnetic spectrum. The science payload of the 5,850-pound (2,660-kg) satellite featured the high-resolution gamma-ray spectrometer experiment (HRGRS), the largest germanium (Ge) spectrometer placed in orbit up to that date. The HRGRS instrument detected energetic photons with energies ranging from 50 keV to 10 MeV, had a field of view of 30 degrees, and had an effective detection area of 11.6 square inches (75 cm²) for 100 keV photons. Depletion of the cryogen for the Ge detectors in the HRGRS took place in mid-1980, effectively ending the major scientific activity of the mission. The spacecraft also performed a cosmic-ray isotope experiment and a heavy nuclei experiment. NASA formally ended the HEAO-3 mission on May 29, 1981. The major scientific accomplishment of *HEAO-3* was a sky survey of gamma-ray sources. This survey expanded activities in the fledgling discipline of gamma-ray astronomy and prepared astronomers and astrophysicists for the *Compton Gamma Ray Observatory,* which operated in the 1990s.

✦ Role of Modern Astrophysics

Today astrophysics has within its reach the ability to bring about one of the greatest scientific achievements ever—a unified understanding of the total evolutionary scheme of the universe. This remarkable revolution in astrophysics is happening now as a result of the confluence of two streams of technical development: remote sensing and spaceflight. Through the science of remote sensing, scientists have acquired sensitive instruments capable of detecting and analyzing radiation across the whole range of the electromagnetic (EM) spectrum. Spaceflight lets astrophysicists place sophisticated remote-sensing instruments above Earth's atmosphere.

The wavelengths transmitted through the interstellar medium and arriving in the vicinity of near-Earth space are spread over approximately 24 decades of the spectrum. (A decade is a group, series, or power of 10.) However, the majority of this interesting EM radiation never reaches the surface of Earth because the terrestrial atmosphere effectively blocks such radiation across most of the spectrum. It should be remembered that the visible and infrared atmospheric windows occupy a spectral slice whose

NASA's COPERNICUS SPACECRAFT

The *Copernicus* spacecraft was launched by NASA on August 21, 1972. This mission was the third in the Orbiting Astronomical Observatory (OAO) program and the second successful spacecraft to observe the celestial sphere from above Earth's atmosphere. An ultraviolet telescope with a spectrometer measured high-resolution spectra of stars, galaxies, and planets, with the main emphasis being placed on the determination of interstellar absorption lines.

Three X-ray telescopes and a collimated proportional counter provided measurements of celestial X-ray sources and interstellar absorption between 1 and 100 angstroms (0.1 to 10 nm) wavelength. Also called the *Orbiting Astronomical Observatory 3*, its observational mission life extended from August 1972 through February 1981—some nine-and-a-half years. NASA named this orbiting observatory in honor of the famous Polish astronomer Nicholas Copernicus.

width is roughly one decade. Ground-based radio observatories can detect stellar radiation over a spectral range that adds about five more decades to the range of observable frequencies, but the remaining 18 decades of the spectrum are still blocked and are effectively invisible to astrophysicists on Earth's surface. Consequently, information that can be gathered by observers at the bottom of Earth's atmosphere represents only a small fraction of the total amount of information available concerning extraterrestrial objects. Sophisticated remote-sensing instruments placed above Earth's atmosphere are now capable of sensing electromagnetic radiation over nearly the entire spectrum, and these instruments are rapidly changing the picture of the cosmos.

Scientists had thought that the interstellar medium was a fairly uniform collection of gas and dust, but spaceborne ultraviolet telescopes have shown them that its structure is very inhomogeneous and complex. There are newly discovered components of the interstellar medium, such as extremely hot gas that is probably heated by shock waves from exploding stars. In fact there is a great deal of interstellar pushing and shoving going on. Matter gathers and cools in some places because matter elsewhere is heated and dispersed. Besides discovering the existence of the very hot gas, the orbiting astronomical observatories have discovered two potential sources of the gas: the intense stellar winds that boil off hot stars and the rarer, but more violent, blasts of matter from exploding supernovas.

In addition, X-ray and gamma-ray astronomy have contributed substantially to the discovery that the universe is not as relatively serene and unchanging as previously imagined, but is actually dominated by the routine occurrence of incredibly violent events. And this series of

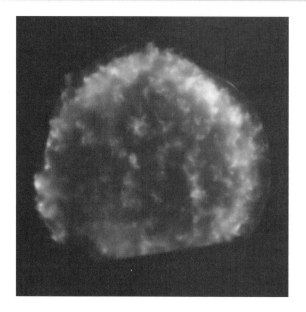

This X-ray image, collected in September 2000 by the *Chandra X-ray Observatory* (*CXO*), reveals fascinating details of the turbulent debris created by the supernova explosion that was observed by the Danish astronomer Tycho Brahe in 1572. To make normally invisible X-rays visible to the human eye, scientists use a false-coloring technique in which colors such as red, green, and blue represent the low–, medium–, and high–energy X-rays detected by the *CXO*. The image is cut off at the bottom because the southernmost region of the supernova remnant fell outside the field of view of the detector. (*NASA/CXC/SAO*)

remarkable new discoveries is just beginning. Future astrophysics missions will provide access to the full range of the electromagnetic spectrum at increased angular and spectral resolution. They will support experimentation in key areas of physics, especially relativity and gravitational physics. Out of these exciting discoveries, perhaps, will emerge the scientific pillars for constructing solar system–level civilization based on technologies unimaginable in the framework of contemporary physics.

Virtually all the information scientists receive about celestial objects comes to them through observation of electromagnetic radiation. Cosmic-ray particles are an obvious and important exception, as are extraterrestrial material samples that have been returned to Earth (for example, lunar rocks or the material in a comet's coma). Each portion of the electromagnetic spectrum carries unique information about the physical conditions and processes in the universe. Infrared radiation reveals the presence of thermal emission from relatively cool objects; ultraviolet and extreme ultraviolet radiation may indicate thermal emission from very hot objects. Various types of violent events can lead to the production of X-rays and gamma rays.

Although EM radiation varies over many decades of energy and wavelength, the basic principles of measurement are quite common to all regions of the spectrum. The fundamental techniques used in astrophysics can be classified as imaging, spectrometry, photometry, and polarimetry. Imaging provides basic information about the distribution of material in a celestial object, its overall structure, and, in some cases, its physical nature. Spectrometry is a measure of radiation intensity as a function of wavelength. It provides information on nuclear, atomic, and molecular phenomena occurring in and around the extraterrestrial object under observation. Photometry involves measuring radiation intensity as a function of time. It provides information about the time variations of physical processes within and around celestial objects, as well as their absolute intensities. Finally, polarimetry is a measurement of radiation intensity as a function of polarization angle. It provides information on ionized particles rotating in strong magnetic fields.

X-ray
Chandra X-ray Observatory

Visible
Hubble Space Telescope

Infrared
Spitzer Space Telescope

Kepler's Supernova Remnant • SN 1604   ssc2004-15b
NASA, ESA / JPL-Caltech / R. Sankrit & W. Blair (Johns Hopkins University)

These three images represent views of Kepler's supernova remnant taken in X-rays, visible light, and infrared radiation. Each top panel shows the entire remnant, while the bottom panels provided close-up views of the remnant. The images show that the bubble of gas that makes up the supernova a remnant appears quite differently in different portions of the electromagnetic spectrum. The *Chandra X-ray Observatory* reveals the hottest gas, which radiates in X-rays. The *Hubble Space Telescope* shows the brightest, most dense gas, which appears in visible light. Finally, the *Spitzer Space Telescope* unveils heated dust, which radiates in infrared light. Since the human eye cannot see X-rays or infrared radiation, astronomers false color-code these data so that they form observable images. *(NASA/ESA/R. Sankrit and W. Blair [John Hopkins University])*

High-energy astrophysics encompasses the study of extraterrestrial X-rays, gamma rays, and energetic cosmic-ray particles. Prior to space-based high-energy astrophysics, scientists believed that violent processes involving high-energy emissions were rare in stellar and galactic evolution. Now, because of studies of extraterrestrial X-rays and gamma rays, they know that such processes are quite common rather than exceptional. The observation of X-ray emissions has been very valuable in the study of high-energy events, such as mass transfer in binary star systems, inter-action of supernova remnants with interstellar gas, and quasars (whose energy source is presently unknown but is believed to involve matter

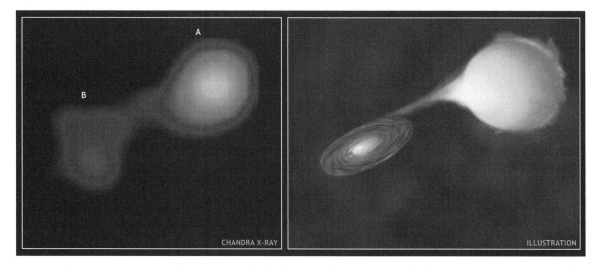

The X-ray image shows the star Mira A (right), a highly evolved red giant, and Mira B (left), a white dwarf that is nibbling on its stellar companion. The artist's rendering on the right depicts the mass transfer that is taking place in the Mira binary star system. Mira A is losing gas rapidly from its upper atmosphere via a stellar wind. Mira B exerts a gravitational tug that creates a gaseous bridge between the stars. High-temperature gas from the wind and bridge accumulates in an accretion disk around Mira B, and collisions between rapidly moving particles in the disk produce X-rays. The separation of the X-rays emanating from the red giant star (Mira A) and the white dwarf (Mira B) was made possible by the superb angular resolution of NASA's *Chandra X-ray Observatory* and the relative proximity of the binary star system, which is about 420 light-years from Earth. *(X-ray image to NASA/ CXC/SAO/M. Karovska et al; artist's rendering to CXC/M. Weiss.)*

accreting [falling into] a black hole). It is thought that gamma rays might be the missing link in understanding the physics of interesting high-energy objects such as pulsars and black holes. The study of cosmic-ray particles provides important information about the physics of nucleosynthesis and about the interactions of particles and strong magnetic fields. High-energy phenomena that are suspected sources of cosmic rays include supernovas, pulsars, radio galaxies, and quasars.

X-ray astronomy is the most advanced of the three high-energy astrophysics disciplines. Space-based X-ray observatories, such as NASA's *Chandra X-ray Observatory,* increase scientific understanding in the following areas: (1) stellar structure and evolution, including binary star systems, supernova remnants, pulsar and plasma effects, and relativity effects in intense gravitational fields; (2) large-scale galactic phenomena, including interstellar media and soft X-ray mapping of local galaxies; (3) the nature of active galaxies, including spectral characteristics and the time variation of X-ray emissions from the nuclear or central regions of such galaxies; and (4) rich clusters of galaxies, including X-ray background radiation and cosmology modeling.

Gamma rays consist of extremely energetic photons—that is, energies greater than 10^5 eV—and result from physical processes different than those associated with X-rays. The processes associated with gamma-ray emissions in astrophysics include: (1) the decay of radioactive nuclei; (2) cosmic-ray interactions; (3) curvature radiation in extremely strong magnetic fields; and (4) matter-antimatter annihilation. Gamma-ray astronomy reveals the explosive, high-energy processes associated with such celestial phenomena as supernovas, exploding galaxies and quasars, and pulsars and black holes.

Observations of iron atoms in the hot gas orbiting three stellar black holes have allowed astronomers to investigate the gravitational effects and spin of these black holes. Data collected by NASA's *Chandra X-ray Observatory* and the European Space Agency's *XMM-Newton* spacecraft suggest that the gravity of a spinning black hole shifts X-ray signals from (accretion cloud) iron atoms to lower energies, producing the strongly skewed X-ray signal.

The orbit of a particle near a black hole depends on the curvature of space around the black hole, which also depends on how fast the black hole is spinning. A spinning black hole drags space around with it and allows atoms to orbit nearer to the black hole than is possible for a non-spinning black hole. The tighter orbit means stronger gravitational effects, which in turn means that more of the X-rays from iron atoms are shifted to lower energies. The most detailed studies of stellar black holes to date indicate that not all black holes spin at the same rate.

Chandra X-ray Observatory data on Cygnus X-1 provide evidence of strong gravitational effects, with some atoms as close as 100 miles (160 km) to the black hole's event horizon, but provide no evidence of spin. *XMM-Newton* data from XTE J1650-500 indicate that some X-rays are coming from as close as 20 miles (32 km) to the black hole's event horizon. Scientists speculate that this black hole must be spinning rapidly. *Chandra X-ray*

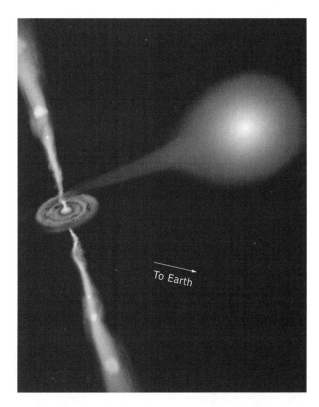

An artist's rendering of SS 433—a binary star system within the Milky Way Galaxy in the constellation Aquila (The eagle)—about 16,000 light-years away. The black hole and its stellar companion are about two-thirds closer to each other than the planet Mercury is to the Sun. The high-speed jets shoot off at 175 million miles (282 million km) per hour—approximately 26 percent of the speed of light. Scientists are using *Chandra X-ray Observatory* data and other observations to help solve the mystery of the great cosmic contradiction, in which black holes, notorious for pulling matter in, somehow manage to also shoot matter away in particle jets moving close to the speed of light. *(CXC/M. Weiss)*

Observatory data of a third stellar black hole, GX 339-4, indicate that this black hole is also spinning rapidly.

One possible explanation for the differences in spin among stellar black holes is that they are born spinning at different rates. Another explanation is that the gas flowing into the black hole spins it up. The black holes with relatively long-lived, low-mass companions, such as XTE J1650-500 and GX 339-4, would have a longer time to spin up than those with massive, short-lived companion stars, such as Cygnus X-1. As space-based X-ray observatories allow scientists to evaluate the spins of additional stellar black holes, it should be possible to test and verify these explanations.

Gamma-ray astronomy is especially significant because the gamma rays being observed can travel across the entire galaxy, and perhaps even across most of the universe, without suffering appreciable alteration or absorption. So scientists postulate that these energetic gamma rays reach the solar system with the same characteristics, including directional and temporal features, as they started with at their sources, possibly many thousands of light-years distant and deep within regions or celestial objects opaque to other wavelengths. To the astrophysicist, gamma-ray astronomy provides information on extraterrestrial phenomena not observable at any other wavelength in the electromagnetic spectrum and on spectacularly energetic events that may have occurred far back in the evolutionary history of the universe.

Cosmic rays are extremely energetic particles that extend in energy from 1 million (10^6) electron volts to over 10^{20} eV and range in composition from hydrogen (atomic number Z = 1) to a predicted atomic number

The artist's rendering shows a nonrotating black hole (on the left) and a rotating black hole (on the right). The most detailed studies of stellar black holes to date indicate that not all black holes spin at the same rate. *(NASA/CXC/M. Weiss)*

of Z = 114. This composition also includes small percentages of electrons, positrons, and possibly antiprotons. Cosmic-ray astronomy provides information on the origin of the elements (nucleosynthetic processes) and the physics of particles at ultra-high-energy levels. Such information addresses astrophysical questions concerning the nature of stellar explosions and the effects of cosmic rays on star formation and galactic structure and stability.

Astronomical work in a number of areas has greatly benefited from large, high-resolution optical systems that have operated or are now operating outside Earth's atmosphere. Some of these interesting areas include investigation of the interstellar medium, detailed study of quasars and black holes, observation of binary X-ray sources and accretion disks, extragalactic astronomy, and observational cosmology. The *Hubble Space Telescope* (*HST*) constitutes the very heart of NASA's contemporary space-borne ultraviolet/optical astronomy program. Launched in 1990, and repaired and refurbished on orbit by several space shuttle missions (discussed in chapter 4), the *HST*'s continued ability to cover a wide range of wavelengths, to provide fine angular resolution, and to detect faint sources makes it one of the most powerful and important astronomical instruments ever built. Within a decade, the *James Webb Space Telescope* will begin operation and provide scientists with even more amazing views of the universe.

Another interesting area of astrophysics involves the extreme ultraviolet (EUV) region of the electromagnetic spectrum. The interstellar medium is highly absorbent at EUV wavelengths (100 to 1,000 angstroms [10 to 100 nm]). EUV data gathered from space-based instruments, such as those placed on NASA's *Extreme Ultraviolet Explorer,* are being used to confirm and refine contemporary theories of the late stages of stellar evolution, to analyze the effects of EUV radiation on the interstellar medium, and to map the distribution of matter in the solar neighborhood.

Infrared astronomy involves studies of the electromagnetic spectrum from one to 100 micrometers wavelength, while radio astronomy involves wavelengths greater than 100 micrometers. Infrared radiation is emitted by all classes of cool objects (stars, planets, ionized gas and dust regions, and galaxies) and the cosmic background radiation. Most emissions from objects with temperatures ranging from about three to 2,000 kelvins are in the infrared region of the spectrum. In order of decreasing wavelength, the sources of infrared and microwave (radio) radiation are: (1) galactic synchrotron radiation; (2) galactic thermal bremsstrahlung radiation in regions of ionized hydrogen; (3) the cosmic background radiation; (4) 15-kelvin cool galactic dust and 100-kelvin stellar-heated galactic dust; (5) infrared galaxies and primeval galaxies; (6) 300-kelvin interplanetary dust; and (7) 3,000-kelvin starlight. Advanced space-based infrared

The majestic dusty spiral galaxy NGC 4414, as imaged by NASA's *Hubble Space Telescope* in a series of observations in 1995 and 1999. This galaxy is about 60 million light-years away. *(NASA, Hubble Heritage Team, AURA, and STScI)*

observatories, such as NASA's *Spitzer Space Telescope,* are revolutionizing how astrophysicists and astronomers perceive the universe.

The *Spitzer Space Telescope* is the fourth and final element in NASA's family of orbiting Great Observatories, which includes the *Hubble Space Telescope,* the *Compton Gamma Ray Observatory,* and the *Chandra X-ray Observatory.* The *Spitzer Space Telescope* joined this spectacular suite of orbiting instruments in August 2003. The sophisticated infrared observatory provides a fresh vantage point on processes that up to now have remained mostly in the dark, such as the formation of galaxies, stars, and planets. Within the Milky Way Galaxy, infrared astronomy is helping scientists discover and characterize dust disks around nearby stars, thought to be the signposts of planetary system formation. The *Spitzer Space Telescope* is also providing new insights into the birth processes of stars—an event normally hidden behind veils of cosmic dust. Outside the galaxy, infrared astronomy helps astronomers and astrophysicists uncover the cosmic engines—thought to be galaxy collisions or black holes—powering ultraluminous infrared galaxies. Infrared astronomy also probes the birth and evolution of galaxies in the early and distant universe.

Gravitation is the dominant long-range force in the universe. It governs the large-scale evolution of the universe and plays a major role in the violent events associated with star formation and collapse. Outer space provides the low acceleration and low-noise environment needed for the careful measurement of relativistic gravitational effects. A number of interesting experiments have been identified for a space-based experimental program in relativity and gravitational physics. Data from these experiments could substantially revise contemporary physics and the fate astrophysicists and cosmologists now postulate for the universe.

The ultimate aim of astrophysics is to understand the origin, nature, and evolution of the universe. It has been said that the universe is not only stranger than people imagine but also stranger than people can ever hope to imagine. Through the creative use of modern space-based platforms, high-energy astronomers and astrophysicists will continue to enjoy many exciting discoveries in astrophysics in the decades ahead. Each new discovery helps human beings understand a little better the magnificent universe in which they live.

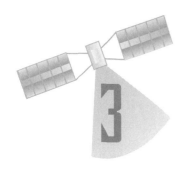

A Revolution in Planetary Astronomy

Robot spacecraft have opened up the universe to exploration. Modern space robots are sophisticated exploring machines that have visited all the major worlds of the solar system, except for tiny Pluto. Emerging out of the politically charged space race of the cold war, a progressively more capable family of robot spacecraft have dramatically changed what scientists know about the alien worlds that journey together with Earth around a star called the Sun.

In a little more than four decades, scientists have learned more about these wandering lights, called πλανετες (planetes, or planets) by the ancient Greek astronomers, than in the previous history of astronomy. Thanks to space robots, every major planetary body and (where appropriate) its collection of companion moons have now become a much more familiar world. Similarly, sophisticated robot astronomical observatories placed on strategically located platforms in space have allowed astronomers and astrophysicists to meet the universe face-to-face, across all the information-rich portions of the electromagnetic spectrum. No longer is the human view of the universe limited to a few narrow bands of radiation that trickle down to Earth's surface through an intervening atmosphere that is often murky and turbulent.

In this century, an armada of ever more sophisticated machine explorers will continue this legacy of exploration as they travel to the farthest reaches of the solar system and beyond. Robot spacecraft have formed a special intellectual partnership with their human creators by allowing us to explore more new worlds in one human lifetime than in the entire history of the human race. This unprecedented wave of discovery and the continued acquisition of vast quantities of new scientific knowledge—perhaps even the first definitive evidence of whether alien life exists or ever existed beyond the boundaries of Earth—will

Robot spacecraft have revolutionized knowledge about the solar system and visited all the major planets (there is an ongoing mission to tiny Pluto). This is a montage of planetary images taken by NASA spacecraft. Included are (from top to bottom) Mercury, Venus, Earth (and Moon), Mars, Jupiter, Saturn, Uranus, and Neptune. The inner planetary bodies (Mercury, Venus, Earth, Moon, and Mars) are roughly to scale to each other; the outer planets (Jupiter, Saturn, Uranus, and Neptune) are also roughly to scale to each other. *(NASA/JPL)*

transform how human beings view themselves and their role in the universe.

There is an interesting correlation between dramatic progress in space exploration and planetary astronomy by robots and parallel spectacular progress in computer technology and aerospace technology. To emphasize the connection, this chapter provides a brief look at some of the most interesting American space robots, as found in the Pioneer, Ranger, Mariner, Viking, and Voyager programs. The main objective is to provide an historic perspective of how space robots emerged from simple, often unreliable, electromechanical exploring devices, into the fairly sophisticated scientific platforms that have extended direct inquiry to the edges of the solar system and created a revolution in planetary astronomy. Several of the most recent and exciting missions to the smaller bodies of the solar system, such as Deep Impact and Stardust, are also included.

✧ A Golden Age of Solar System Exploration

The American space age began on January 31, 1958, with the launch of the first U.S. satellite—*Explorer 1*—an Earth-orbiting spacecraft built and controlled by the Jet Propulsion Laboratory (JPL). *Explorer 1* discovered Earth's trapped radiation belts and heralded the great wave of scientific discovery that would become an integral part of the Space Age. For almost five decades since the launch of *Explorer 1,* the JPL has led the world in exploring the solar system with robot spacecraft.

The Jet Propulsion Laboratory is a federally funded research and development facility managed by the California Institute of Technology for the National Aeronautics and Space Administration (NASA). The laboratory is located in Pasadena, California, approximately 20 miles (32 km) northeast of Los Angeles. In addition to the Pasadena site, the JPL operates the worldwide Deep Space Network (DSN), including a DSN station, at Goldstone, California.

The JPL's origin dates back to the 1930s, when Caltech professor Theodore von Kármán (1881–1963) supervised pioneering work in rocket propulsion for the U.S. Army—including the use of strap-on rockets for "jet-assisted take-off" of aircraft with extra-heavy cargoes. At the time, von Kármán was head of Caltech's Guggenheim Aeronautical Laboratory. On December 3, 1958, two months after the U.S. Congress created NASA, the JPL was transferred from U.S. Army jurisdiction to that of the new civilian space agency. The laboratory now covers 177 acres (72 ha) adjacent to the site of von Kármán's early rocket experiments in a dry riverbed wilderness area of Arroyo Seco.

In the 1960s, the JPL began to conceive, design, and operate robot spacecraft to explore other worlds. This effort initially focused on NASA's Ranger and Surveyor Moon missions— robot spacecraft that paved the way for successful human landings by the Apollo Project astronauts. The Ranger spacecraft were the first U.S. robot spacecraft sent toward the Moon in the early 1960s to prepare the way for the Apollo Project's human landings at the end of that decade. The Rangers were a series of fully attitude-controlled robot spacecraft designed to photograph the lunar surface at close range before impacting. *Ranger 1* was launched on August 23, 1961, from Cape Canaveral Air Force Station and set the stage for the rest of the Ranger missions by testing spacecraft navigational performance. The *Ranger 2* through *9* spacecraft were launched from November 1961 through March 1965. All of the early Ranger missions (namely, *Ranger 1* through *6*) suffered setbacks of one type or another. Finally, *Ranger 7, 8,* and *9* succeeded with flights that returned many thousands of lunar surface images (before impact) and greatly advanced scientific knowledge about the Moon.

A photograph of the Moon's Crater Alphonsus taken by NASA's *Ranger 9* spacecraft at an altitude of 265 miles (426 km) on July 31, 1964. Three minutes after this picture was taken, the space probe impacted on the lunar surface (in area marked by white circle). *(NASA/JPL)*

NASA's highly successful Surveyor Project began in 1960. It consisted of seven robot-lander spacecraft that were launched between May 1966 and January 1968, as an immediate precursor to the human expeditions to the lunar surface in the Apollo Project. These versatile space robots were used to develop soft-landing techniques, to survey potential Apollo mission landing sites, and to improve scientific understanding of the Moon.

The *Surveyor 1* spacecraft was launched on May 30, 1966, and soft-landed in the Ocean of Storms region of the Moon. The space robot discovered that the bearing strength of the lunar soil was more than adequate to support the Apollo Project's human-crewed, lander spacecraft (called the lunar excursion module, or LEM). This finding contradicted the then-prevalent hypothesis that a heavy spacecraft like the LEM might sink out of sight in the anticipated talcum powder-like, ultrafine lunar dust particles. The *Surveyor 1* spacecraft also telecast many images from the lunar surface.

Surveyor 2 was the second in this series of soft-landing robots. Successfully launched on September 20, 1966, by an Atlas-Centaur rocket from Cape Canaveral, this robot lander experienced a vernier engine failure during a midcourse maneuver while en route to the Moon. The failure of one vernier engine to fire resulted in an unbalanced thrust that caused *Surveyor 2* to tumble. Attempts by NASA engineers to salvage this mission failed.

Things went much better for NASA's next robot-lander mission to the Moon. The *Surveyor 3* spacecraft was launched on April 17, 1967, and soft-landed on the side of a small crater in another region of the Ocean of Storms. The perky space robot used the shovel attached to its mechanical arm to dig a trench and discovered that the load-bearing strength of the lunar soil increased with depth. *Surveyor 3* also transmitted many images from the lunar surface.

At the same time that JPL engineers were busy with the Ranger and Surveyor missions, they also conducted Mariner spacecraft missions to

An artist's rendering of NASA's *Mariner 2* spacecraft—the world's first successful interplanetary spacecraft. Launched on August 27, 1962, from Cape Canaveral, *Mariner 2* flew past the planet Venus on December 14, 1962, at an encounter distance of about 25,480 miles (41,000 km). *(NASA/JPL)*

Mercury, Venus, and Mars. The Mariner missions were truly trailblazing efforts that continued through the early 1970s and greatly revised scientific understanding of the terrestrial planets and the inner solar system. The first Mariner mission, called *Mariner 1,* was intended to perform a Venus flyby. NASA and JPL engineers based the design of this spacecraft on the Ranger lunar spacecraft. A successful liftoff of *Mariner 1*'s Atlas-Agena B launch vehicle on July 22, 1962, soon turned tragic. When the rocket vehicle veered off course, the range safety officer at Cape Canaveral Air Force Station was forced to destroy it some 293 seconds after launch. Because of faulty guidance commands, the rocket vehicle's steering was very erratic and the *Mariner 1* spacecraft was going to crash somewhere on Earth, possibly in the North Atlantic shipping lanes or in an inhabited area. Undaunted by the heartbreaking loss of the *Mariner 1* spacecraft, which was never given a chance to demonstrate its capabilities, the NASA/JPL engineering team quickly prepared its identical twin, named *Mariner 2,* to pinch-hit and perform the world's first interplanetary flyby mission.

Following a successful launch from Cape Canaveral on August 27, 1962, *Mariner 2* cruised through interplanetary space, and then became the first robot spacecraft to fly past another planet (in this case, Venus). *Mariner 2* encountered Venus at a distance of about 25,500 miles (41,000 km) on December 14, 1962. Following the flyby of Venus, *Mariner 2* went into orbit around the Sun. The scientific discoveries made by *Mariner 2* included a slow retrograde rotation rate for Venus, hot surface temperatures and high surface pressures, a predominantly carbon dioxide atmosphere, continuous cloud cover with a top altitude of about 37 miles (60 km), and no detectable magnetic field. Data collected by *Mariner 2* during its interplanetary journey to Venus showed that the solar wind streams continuously in interplanetary space and that the cosmic dust density is much lower than in the region of space near Earth.

The *Mariner 2* encounter helped scientists dispel many pre–space age romantic fantasies about Venus, including the widely held speculation (which appeared in both the science and science-fiction literature) that the cloud-shrouded planet was a prehistoric world, mirroring a younger Earth. Except for a few physical similarities like size and surface gravity level, robot spacecraft visits in the 1960s and 1970s continued to show that Earth and Venus were very different worlds. For example, the surface temperature on Venus reaches almost 932°F (500°C), its atmospheric pressure is more than 90 times that of Earth, it has no surface water, and its dense atmosphere with sulfuric acid clouds and an overabundance of carbon dioxide (about 96 percent) represents a runaway greenhouse of disastrous proportions.

The next Mariner project undertaken by NASA and the JPL targeted the planet Mars. Two spacecraft were prepared, *Mariner 3* and its backup,

VENERA PROBES AND SPACECRAFT

The Venera probes and spacecraft were a family of mostly successful robotic space missions flown by the Soviet Union to the planet Venus between 1961 and 1984. These missions included orbiters, landers, and atmospheric probes. In October 1967, the *Venera 4* spacecraft placed a landing capsule/probe into the Venusian atmosphere, collecting data that indicated the planet's atmosphere was from 90 to 95 percent carbon dioxide (CO_2). This 840-pound (380-kg) probe descended by parachute for about 94 minutes, when data transmissions ceased at an altitude of some 15.5 miles (25 km).

In December 1970, the 1,090-pound (495-kg) atmospheric probe from the *Venera 7* spacecraft reached the surface of Venus and transmitted data for about 20 minutes. The *Venera 8* spacecraft launched a probe into the Venusian atmosphere in July 1972. This probe, with an improved communications system, successfully landed on the surface and survived for about 50 minutes in the inferno-like conditions found on the surface of Venus.

In October 1975, the *Venera 9* and *10* spacecraft sent probes to the surface of Venus that landed successfully and transmitted the first black-and-white images of the planet's rock-strewn surface. *Venera 11* and *12*, launched in 1978, were also lander probe/flyby missions with improved sensors.

In the 1980s, Soviet space scientists launched four more sophisticated Venera spacecraft to Venus. The *Venera 13* and *14* spacecraft were 11,000-pound (5,000-kg) flyby/lander configurations sent to the planet at the end of 1981. *Venera 13* landed on Venus on March 1, 1982, and its identical companion, *Venera 14*, touched down on March 5, 1982. These landers returned black-and-white and

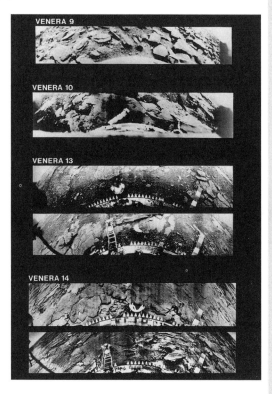

Wide-angle views of the surface of Venus from the Soviet (Russian) *Venera 9, 10, 13,* and *14* lander spacecraft *(NASA)*

color images of Venus's surface. They also performed the first soil analysis of the Venusian surface.

The final pair of Venera spacecraft (called *Venera 15* and *16*), were launched in June 1983. These spacecraft did not carry landers, but rather performed mapping missions of the planet's cloud-enshrouded surface, using their synthetic aperture radar systems. The spacecraft orbited for about one year and produced detailed radar images of the planet's surface at a resolution of between 0.6 and 1.2 miles (1 and 2 km).

Mariner 4 (an identical twin). *Mariner 3* was launched from Cape Canaveral on November 5, 1964, but the shroud encasing the spacecraft atop its rocket failed to open properly, and *Mariner 3* did not get to Mars. Three weeks later, *Mariner 4* was launched successfully and sent on an eight-month voyage to the Red Planet. How was such a quick recovery and new launch possible in so short a time?

In the early days of space exploration, launch-vehicle failures were quite common, so aerospace engineers and managers considered it prudent to build two (or more) identical spacecraft for each important mission. Should one spacecraft experience a fatal launch accident, the other spacecraft could quickly be readied to take advantage of a particular interplanetary launch window. If both spacecraft proved successful, the scientific return for that particular mission usually more than doubled. In this fortunate case, scientists could use the preliminary findings of the first space robot to guide the data-collection efforts of the second robot, as it approached the target planet several weeks later.

NASA's three most successful robot-twin missions of the 1970s were *Pioneer 10* and *11* (flybys), *Viking 1* and *2* (landers and orbiters), and *Voyager 1* and *2* (flybys). Starting in 2004, fortune smiled again when NASA's twin Mars Exploration Rovers (MERs), named *Spirit* and *Opportunity*, arrived safely on the Red Planet and began moving across the surface at the start of highly productive scientific investigations.

A launch window is the time interval during which a spacecraft can be sent to its destination. An interplanetary launch window is generally constrained to within a few weeks each year (or less) by the location of Earth in its orbit around the Sun. Proper timing allows the launch vehicle to use Earth's orbital motion in its overall trajectory. Earth departure timing is also critical if the spacecraft is to arrive at a particular point in interplanetary space just as the target planet does. By carefully choosing the launch window, interplanetary spacecraft can employ a minimum energy path called the Hohmann transfer trajectory, after the German engineer Walter Hohmann (1880–1945), who described this orbital transfer technique in 1925. Orbital mechanics, payload mass, and rocket-vehicle thrust all influence interplanetary travel.

The most energy-efficient launch windows from Earth to Mars occur about every two years. Determining launch windows for missions to the giant outer planets proves a bit more complicated. For example, only once every 176 years do the four giant planets (Jupiter, Saturn, Uranus, and Neptune) align themselves in such a pattern that a spacecraft launched from Earth to Jupiter at just the right time might be able to visit the three other giant planets on the same mission, using a technique called gravity assist. This unique opportunity occurred in 1977, and NASA scientists took advantage of a special celestial alignment by launching two sophisticated robot spacecraft, called *Voyager 1* and *2*, on multiple giant planet encounter missions. As described shortly, *Voyager 1* visited Jupiter and Saturn, while *Voyager 2* took the so-called grand tour and visited all four giant planets on the same mission.

In the cold-war environment of the early 1960s, a great deal of political emphasis and global attention was given to achievements in space exploration. The superpower that accomplished this or that space exploration first earned a central position on the world political stage. So, NASA managers soon recognized that building identical-twin spacecraft (just in case one did not complete the mission) proved a relatively inexpensive approach to pursuing major scientific objectives while earning political capital. Superpower competition during the cold war fueled an explosion in space exploration and produced an age of discovery unprecedented in history. Primarily due to robot spacecraft, more scientific information about the solar system and the universe was collected in the past four decades than in all previous human history. That exciting wave of discovery continues in the post–cold war era, as more sophisticated space robots, such as *Cassini/Huygens,* explore the unknown.

Before discussing the spectacular results of the Viking mission or the great journeys of the Voyager spacecraft, let us return our attention to the very important Mariner 4 mission to Mars. *Mariner 4* was successfully launched from Cape Canaveral on November 28, 1964. It traveled for almost eight months through interplanetary space, and then zipped past Mars on July 14, 1965. At its closest approach, *Mariner 4* was only 6,120 miles (9,845 km) from the surface of Mars during the flyby. As this space robot encountered Mars, it collected the first close-up images of another planet. These images, played back from a small tape recorder over a long period, showed lunar-type impact craters, some of them touched with frost in the chill of the Martian evening. *Mariner 4*'s 21 complete pictures (plus 21 lines of a 22nd picture) might be regarded as quite crude when compared to the high-resolution imagery of Mars provided by contemporary robot spacecraft. However, these first images of another world started a revolution that overturned many long-cherished views about the Red Planet.

Throughout human history, the Red Planet, Mars, has been at the center of astronomical thought. The ancient Babylonians followed the motions of this wandering red light across the night sky and named it after Nergal, their god of war. In time the Romans, also honoring their own god of war, gave the planet its present name. The presence of an atmosphere, polar caps, and changing patterns of light and dark on the surface caused many pre–space age astronomers and scientists to consider Mars an Earth-like planet—the possible abode of extraterrestrial life. The American astronomer Percival Lowell (1855–1916) was one of the most outspoken proponents of the canal theory. In several popular publications, he insisted that Mars was a dying planet whose intelligent inhabitants constructed huge canals to distribute a scarce supply of water around the alien world. Since H. G. Wells published his classic science-fiction story *War of the Worlds* at the end of the 19th century, an invasion from Mars has remained one of the most enduring and popular themes in both the science-fiction literature and in the motion-picture industry.

However, with *Mariner 4* leading the scientific parade, a wave of sophisticated robot spacecraft—flybys, orbiters, landers, and rovers—have shattered the canal theory, that persistent romantic myth of a race of ancient Martians struggling to bring water from the polar caps to the more productive regions of a dying world. Spacecraft-derived data have shown that the Red Planet is actually a halfway world. Part of the Martian surface is ancient, like the surfaces of the Moon and Mercury, while part is more evolved and Earth-like. Mars remains at the center of intense investigation by a new wave of sophisticated robot spacecraft. The continued search for microbial life (existent or extinct) and the resolution of the intriguing mystery about the fate of the liquid water that appears to have flowed on ancient Mars in large quantities top NASA's current space-exploration agenda.

Other successful Mariner missions included *Mariner 5,* launched in 1967 to Venus; *Mariner 6,* launched in 1969 to Mars; *Mariner 7,* launched in 1969 to Mars; and *Mariner 9,* launched in 1971 to Mars. In November 1971, *Mariner 9* became the first artificial satellite of Mars and the first spacecraft of any country to orbit another planet. The robot spacecraft waited patiently for a giant planetwide dust storm to abate, and then it compiled a collection of high-quality images of the surface of Mars that provided scientists with their first global mosaic of the Red Planet. *Mariner 9* also took the first close-up images of the two small Martian satellites, Phobos and Deimos.

Mariner 10 became the first spacecraft to use a gravity-assist boost from one planet to send it to another planet—a key innovation in space-flight that enabled exploration of the outer planets by robot spacecraft. *Mariner 10*'s launch from Cape Canaveral in November 1973 delivered the spacecraft to Venus in February 1974, where a gravity-assist flyby allowed it to encounter the planet Mercury in March and September of that year. *Mariner 10* was the first and, thus far, the only spacecraft of any country to explore the innermost planet in the solar system. On August 3, 2004, NASA launched *MESSENGER* from Cape Canaveral and sent the orbiter spacecraft on a long-interplanetary journey to Mercury. In March 2011, *MESSENGER* is set to become the first robot spacecraft to achieve orbit around Mercury.

A mosaic image of Mercury based on photographs taken by NASA's *Mariner 10* spacecraft during its three encounters with the planet (1974–75) *(NASA/JPL)*

The first intense search for life on Mars was begun in 1975, when NASA launched the agency's Viking missions, consisting of two orbiter and two lander spacecraft. The Viking Project was the culmination of an initial series of American missions to explore Mars in the 1960s and 1970s. This series of interplanetary missions began in 1964 with *Mariner 4,* continuing with the *Mariner 6* and *7* flyby missions in 1969 and then the *Mariner 9* orbital mission in 1971 and 1972.

Viking was designed to orbit Mars and to land and operate on the surface of the Red Planet. Two identical spacecraft, each consisting of a lander and an orbiter, were built. Both Viking missions were launched from Cape Canaveral, Florida. *Viking 1* was launched on August 20, 1975, and *Viking 2* on September 9, 1975. The landers were sterilized before launch to prevent contamination of Mars by terrestrial microorganisms. These spacecraft spent nearly a year in transit to the Red Planet. *Viking 1* achieved Mars orbit on June 19, 1976, and *Viking 2* began orbiting Mars on August 7, 1976. The *Viking 1* lander accomplished the first soft landing on Mars on July 20, 1976, on the western slope of Chryse Planitia (The plains of gold)

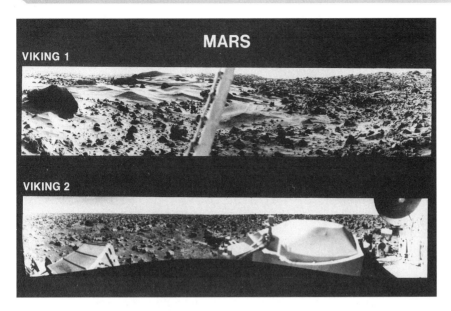

Panoramic views of the surface of Mars from the *Viking 1* and *2* lander spacecraft. The *Viking 1* lander accomplished the first soft landing on Mars on July 20, 1976, on the western slope of Chryse Planitia (The plains of gold) at 22.46 degrees north latitude, 48.01 degrees west longitude. The *Viking 2* lander touched down successfully on September 3, 1976, at Utopia Planitia (The plains of utopia) located at 47.96 degrees north latitude, 225.77 degrees west longitude. *(NASA)*

at 22.46 degrees north latitude, 48.01 degrees west longitude. The *Viking 2* lander touched down successfully on September 3, 1976, at Utopia Planitia located at 47.96 degrees north latitude, 225.77 degrees west longitude.

The Viking mission was planned to continue for 90 days after landing. Each orbiter and lander, however, operated far beyond its design lifetime. For example, the *Viking 1* orbiter exceeded four years of active flight operations in orbit around Mars. The Viking Project's primary mission ended on November 15, 1976, just 11 days before Mars passed behind the Sun (an astronomical event called a superior conjunction). After conjunction, in mid-December 1976, telemetry and command operations were reestablished and extended mission operations began.

The mission of the *Viking 2* orbiter spacecraft ended on July 25, 1978, due to exhaustion of attitude-control system gas. The *Viking 1* orbiter spacecraft also began to run low on attitude-control system gas, but through careful planning, it was possible to continue collecting scientific data (at a reduced level) for another two years. Finally, with its control gas supply exhausted, the *Viking 1* orbiter's electrical power was commanded off on August 7, 1980.

The last data from the *Viking 2* lander were received on April 11, 1980. The *Viking 1* lander made its final transmission on November 11, 1982. After more than six months of effort to regain contact with the *Viking 1* lander, the Viking mission came to an end on May 23, 1983.

With the single exception of the seismic instruments, the entire complement of scientific instruments of the Viking Project acquired far more data about Mars than ever anticipated. The primary objective of the lander was to determine whether (microbial) life currently exists on Mars. The evidence provided by the landers is still subject to debate, although most scientists feel these results are strongly indicative that life does not now exist on Mars. However, recent analyses of Martian meteorites have renewed interest in this very important question, and Mars is once again the target of intense scientific investigation by even more sophisticated scientific spacecraft.

The *Pioneer 10* and *11* spacecraft were designed as true deep-space robot explorers—the first human-made objects to navigate the main asteroid belt, the first spacecraft to encounter Jupiter and its fierce radiation belts, the first to encounter Saturn, and the first spacecraft to leave the solar system. This far-traveling pair of robot spacecraft also investigated magnetic fields, cosmic rays, the solar wind, and the interplanetary dust concentrations as they flew through interplanetary space.

Credit for the single space-robot mission that has visited the most planets goes to the JPL's Voyager project. Launched in 1977, the twin *Voyager 1* and *Voyager 2* spacecraft flew by the planets Jupiter (1979) and Saturn (1980–81). *Voyager 2* then went on to have an encounter with Uranus (1986) and with Neptune (1989). On August 25, 1989, *Voyager 2* encountered Neptune at a closest approach distance of 3,100 miles (5,000 km). Many space historians treat this planetary flyby as the end of the first golden age of planetary exploration brought about by space technology and robot spacecraft.

Both *Voyager 1* and *Voyager 2* are now traveling on different trajectories into interstellar space. In February 1998, *Voyager 1* passed the *Pioneer 10* spacecraft to become the most distant human-made object in space. The Voyager Interstellar Mission should continue well into the next decade.

Millions of years from now—most likely when human civilization has completely disappeared from the surface of Earth—four robot spacecraft (*Pioneer 10* and *11; Voyager 1* and *2*) will continue to drift through the interstellar void. Each spacecraft will serve as a legacy of human ingenuity and inquisitiveness. By carrying a special message from Earth, each far-traveling robot spacecraft also bears permanent testimony that for at least one moment in the history of the human species, a few people raised their foreheads to the sky and reached for the stars. Though primarily designed for scientific inquiry within the solar system, these four relatively simple robotic exploring machines are now a more enduring artifact of human civilization than any cave painting, petroglyph, great monument, giant palace, or high-rising city created here on Earth.

✧ The New Wave of Planetary Exploration

A new generation of more sophisticated spacecraft appeared in the late 1980s and early 1990s. These spacecraft allowed NASA to conduct much more detailed scientific investigations of the planets and the Sun. Representative of the significant advances in sensor technology, computer technology, and aerospace engineering are the robot spacecraft used in the Galileo mission to Jupiter and the Cassini mission to Saturn.

GALILEO MISSION

The Galileo mission began on October 18, 1989, when the sophisticated spacecraft was carried into low Earth orbit by the space shuttle *Atlantis* and then started on its interplanetary journey by means of an inertial upper stage rocket. Relying on gravity-assist flybys to reach Jupiter, the *Galileo* spacecraft flew past Venus once and Earth twice. As it traveled through interplanetary space beyond Mars on its way to Jupiter, *Galileo* encountered the asteroids Gaspra (October 1991) and Ida (August 1993). *Galileo's* flyby of Gaspra on October 29, 1991, gave scientists their first-ever close-up look at a minor planet. On its final approach to Jupiter, *Galileo* observed the giant planet being bombarded by fragments of Comet Shoemaker-Levy-9, which had broken apart. On July 12, 1995, the *Galileo* mother spacecraft separated from its hitchhiking companion (an atmospheric probe), and the two robot spacecraft flew in formation to their final destination.

On December 7, 1995, *Galileo* fired its main engine to enter orbit around Jupiter and gathered data transmitted from the atmospheric probe during that small robot's parachute-assisted descent into the Jovian atmosphere. During its two-year prime mission, the *Galileo* spacecraft performed 10 targeted flybys of Jupiter's major moons. In December 1997, the sophisticated robot spacecraft began an extended scientific mission that featured eight flybys of Jupiter's smooth, ice-covered moon Europa and two flybys of its pizza-colored, volcanic moon Io.

Galileo started a second extended scientific mission in early 2000. This mission included flybys of the Galilean moons Io, Ganymede, and Callisto, plus coordinated observations of Jupiter with the *Cassini* spacecraft. In December 2000, *Cassini* flew past the giant planet to receive a much-needed gravity assist that enabled the large spacecraft to eventually reach

Saturn. *Galileo* conducted its final flyby of a Jovian moon in November 2002, when it zipped past the tiny inner moon Amalthea.

The encounter with Amalthea left *Galileo* on a course that would lead to an intentional impact into Jupiter in September 2003. NASA mission controllers deliberately crashed the *Galileo* mother spacecraft into Jupiter at the end of the space robot's very productive scientific mission to avoid any possibility of contaminating Europa with terrestrial microorganisms. As a uncontrolled derelict, the *Galileo* spacecraft might have eventually crashed into Europa sometime within the next few decades. Many exobiologists suspect that Europa has a life-bearing, liquid-water ocean underneath its icy surface. Since the *Galileo* spacecraft was probably harboring a variety of hitchhiking terrestrial microorganisms, scientists thought it prudent to completely avoid any possibility of contamination of Europa. The easiest way of resolving the potential problem was to simply dispose of the retired *Galileo* spacecraft in the frigid, swirling clouds of Jupiter.

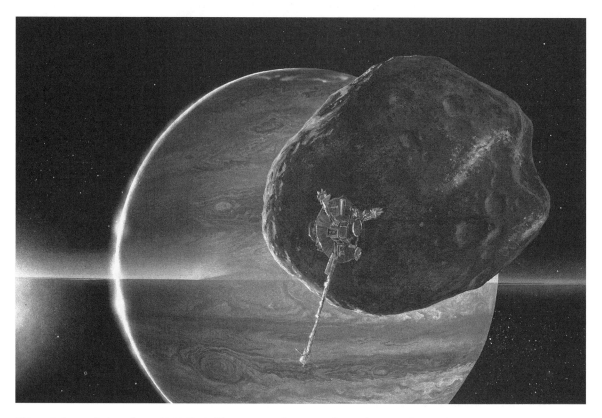

This artist's rendering shows NASA's *Galileo* spacecraft as it performed a very close flyby of Jupiter's tiny inner moon, Amalthea, in November 2002. *(NASA)*

So, NASA and JPL mission controllers accomplished this task, while still maintaining sufficient control over *Galileo's* behavior and trajectory.

CASSINI/HUYGENS SPACECRAFT

The Cassini/Huygens mission was successfully launched by a mighty Titan IV-Centaur vehicle on October 15, 1997, from Cape Canaveral Air Force Station, Florida. It is a joint NASA and European Space Agency (ESA) project to conduct detailed exploration of Saturn, its major moon Titan, and its complex system of other moons. Following the example of the *Galileo* spacecraft, the *Cassini* spacecraft also took a gravity-assisted tour of the solar system. The spacecraft eventually reached Saturn following a Venus-Venus-Earth-Jupiter gravity assist (VVEJGA) trajectory. After a nearly seven-year journey through interplanetary space covering 2.2 billion miles (3.5 billion km), the *Cassini* spacecraft arrived at Saturn on July 1, 2004 (Eastern daylight time).

The very large and complex robot spacecraft is named in honor of Italian-born French astronomer Giovanni Domenico Cassini (1625–1712), who was the first director of the Royal Observatory in Paris and conducted extensive observations of the Saturn. The *Huygens* probe is named in honor of Dutch astronomer Christiaan Huygens (1629–95), who discovered Titan in 1655.

The most critical phase of the mission after launch was Saturn orbit insertion (SOI). When *Cassini* arrived at Saturn, the sophisticated robot spacecraft fired its main engine for 96 minutes to reduce its speed and allow it to be captured as a satellite of Saturn. Passing through a gap between Saturn's F and G rings, the intrepid spacecraft successfully swung close to the planet and began the first of some six-dozen orbits it will complete during its four-year primary mission.

The arrival period provided a unique opportunity to observe Saturn's rings and the planet itself, since this was the closest approach the spacecraft will make to Saturn during the entire mission. As anticipated, the *Cassini* spacecraft went right to work upon arrival and provided scientific results.

Scientists examining Saturn's contorted F ring, which has baffled them since its discovery, have found one small body, possibly two, orbiting in the F-ring region and a ring of material associated with Saturn's moon Atlas. *Cassini's* close-up look at Saturn's rings revealed a small object moving near the outside edge of the F ring, interior to the orbit of Saturn's moon Pandora. This tiny object, which is about 3.1 miles (5 km) in diameter, has been provisionally assigned the name S/2004 S3. It may be a tiny moon that orbits Saturn at a distance of 87,600 miles (141,000 km) from Saturn's center. This object is located about 620 miles (1,000 km)

An artist's concept of the *Cassini* spacecraft during the critical Saturn orbit-insertion (SOI) maneuver, just after the main engines had begun firing on July 1, 2004. The SOI maneuver reduced the robot spacecraft's speed, allowing *Cassini* to be captured by Saturn's gravity and enter orbit. Following the successful SOI maneuver, *Cassini* began a planned four-year exploration mission of Saturn, its mysterious moons, stunning rings, and complex magnetic environment. On December 25, 2004, *Cassini* released its hitchhiking companion, the *Huygens* probe—sending the robot on an historic one-way journey into the atmosphere of Saturn's largest moon, Titan. *(NASA/JPL)*

from Saturn's F ring. A second object, provisionally called S/2004 S4, has also been observed in the initial imagery provided by the *Cassini* spacecraft. About the same size as S/2004 S3, this object appears to exhibit some strange dynamics, which take it across the F ring.

In the process of examining the F-ring region, scientists also detected a previously unknown ring, now called S/2004 1R. This new ring is associated with Saturn's moon Atlas. The ring is located 85,770 miles (138,000 km) from the center of Saturn in the orbit of the moon Atlas, between the A ring and the F ring. Scientists estimate the ring's width at 185 miles (300 km).

Upon arrival at Saturn and its successful orbit-insert burn (July 2004), the *Cassini* spacecraft began its extended tour of the Saturn system. This orbital tour involves at least 76 orbits around Saturn, including 52 close encounters with seven of Saturn's known moons. The *Cassini* spacecraft's orbits around Saturn are being shaped by gravity-assist flybys of Titan. Close flybys of Titan also permit high-resolution mapping of the intriguing, cloud-shrouded moon's surface. The *Cassini* orbiter spacecraft carries an instrument called the Titan imaging radar, which can see through the opaque haze covering that moon to produce vivid topographic maps of the surface.

The size of these orbits, their orientation relative to Saturn and the Sun, and their inclination to Saturn's equator are dictated by various scientific requirements. These scientific requirements include: imaging radar coverage of Titan's surface; flybys of selected icy moons, Saturn, or Titan; occultations of Saturn's rings; and crossings of the ring plane.

The *Cassini* orbiter will make at least six close, targeted flybys of selected icy moons of greatest scientific interest—namely, Iapetus, Enceladus, Dione, and Rhea. Images taken with *Cassini's* high-resolution telescopic cameras during these flybys will show surface features equivalent in spatial resolution to the size of a professional baseball diamond. At least two dozen more distant flybys (at altitudes of up to 62,000 miles [100,000 km]) will also be made of the major moons of Saturn—other than Titan. The varying inclination of the *Cassini* spacecraft's orbits around Saturn will allow the spacecraft to conduct studies of the planet's polar regions, as well as its equatorial zone.

In addition to the *Huygens* probe, Titan will be the subject of close scientific investigations by the *Cassini* orbiter. The spacecraft will execute 45 targeted, close flybys of Titan, Saturn's largest moon. Some flybys will be as close as about 590 miles (950 km) above the surface. Titan is the only Saturn moon large enough to enable significant gravity-assist changes in *Cassini's* orbit. Accurate navigation and targeting of the point at which the *Cassini* orbiter flies

Nine days before it entered orbit around Saturn, NASA's *Cassini* spacecraft captured this exquisite view of Saturn's rings. The image was obtained from a vantage point beneath the ring plane with the narrow-angle camera on June 21, 2004, at a distance of 4 million miles (6.4 million km) from Saturn. *(NASA/JPL)*

by Titan will be used to shape the orbital tour. This mission-planning approach is similar to the way the *Galileo* spacecraft used its encounters of Jupiter's large moons (the Galilean satellites) to shape its very successful scientific tour of the Jovian system.

As currently planned, the prime mission tour of the *Cassini* spacecraft will end on June 30, 2008. This date is four years after arrival at Saturn and 33 days after the last Titan flyby, which will occur on May 28, 2008. The aim point of the final flyby is being chosen to position *Cassini* for an additional Titan flyby on July 31, 2008—providing mission controllers the opportunity to proceed with more flybys during an extended missions, if resources (such as the supply of attitude-control propellant) allow. Nothing in the present design of the orbital tour of the Saturn system now precludes an extended mission.

The *Cassini* spacecraft, which originally included the orbiter and the *Huygens* probe, is the largest and most complex interplanetary spacecraft ever built. The orbiter spacecraft alone has a dry mass of 4,675 pounds (2,125 kg). When the 704-pound (320-kg) *Huygens* probe and a launch-vehicle adapter were attached and 6,890 pounds (3,130 kg) of attitude control and maneuvering propellants loaded, the assembled spacecraft acquired a total launch mass of 12,570 pounds (5,712 kg). At launch the fully assembled *Cassini* spacecraft stood 22-feet- (6.7-m-) high and 12.9-feet- (4-m-) wide.

The Cassini mission involves a total of 18 science instruments, six of which are contained in the wok-shaped *Huygens* probe. This ESA-sponsored probe was detached from the main orbiter spacecraft *Cassini* on December 25, 2004, and successfully conducted its own scientific investigations as it plunged into the atmosphere of Titan on January 14, 2005. The probe's science instruments included: the aerosol collector pyrolyzer, descent imager and spectral radiometer, Doppler wind experiment, gas chromatograph and mass spectrometer, atmospheric structure instrument, and surface science package.

The *Cassini* spacecraft's science instruments include: a composite infrared spectrometer, imaging system, ultraviolet-imaging spectrograph, visual- and infrared-mapping spectrometer, imaging radar, radio science, plasma spectrometer, cosmic-dust analyzer, ion and neutral mass spectrometer, magnetometer, magnetospheric-imaging instrument, and radio- and plasma-wave science. Telemetry from the spacecraft's communications antenna is also being used to make observations of the atmospheres of Titan and Saturn and to measure the gravity fields of the planet and its satellites.

The Cassini mission (including *Huygens* probe and orbiter spacecraft) is designed to perform a detailed scientific study of Saturn, its rings, its magnetosphere, its icy satellites, and its major moon Titan. The *Cassini* orbiter's scientific investigation of the planet Saturn includes: cloud properties and atmospheric composition; winds and temperatures; internal structure and rotation; the characteristics of the ionosphere; and the origin and evolution of the planet. Scientific investigation of the Saturn ring system includes: structure and composition; dynamic processes within the rings; the interrelation of rings and satellites; and the dust and micrometeoroid environment.

Saturn's magnetosphere involves the enormous magnetic bubble surrounding the planet that is generated by its internal magnet. The magnetosphere also consists of the electrically charged and neutral particles within this magnetic bubble. Scientific investigation of Saturn's magnetosphere includes: its current configuration; particle composition, sources and sinks; dynamic processes; its interaction with the solar wind, satellites, and rings; and Titan's interaction with both the magnetosphere and the solar wind.

During the orbit tour phase of the mission (from July 1, 2004 to June 30, 2008), the *Cassini* orbiter spacecraft will perform many flyby encounters of all the known icy moons of Saturn. As a result of these numerous satellite flybys, the spacecraft's instruments will investigate: the characteristics and geologic histories of the icy satellites; the mechanisms for surface modification; surface composition and distribution; bulk composition

These two montages of Saturn's moon Phoebe, taken by NASA's *Cassini* spacecraft in June 2004, show the names provisionally assigned to 24 craters on this satellite by the International Astronomical Union. *(NASA/JPL)*

and internal structure; and interaction of the satellites with Saturn's magnetosphere.

The moons of Saturn are diverse—ranging from the planetlike Titan to tiny, irregular objects only tens of kilometers in diameter. Scientists currently believe that all of these bodies (except perhaps for Phoebe) hold not only water ice but also other chemical components, such as methane, ammonia, and carbon dioxide. Before the advent of robotic spacecraft in space exploration, scientists believed the moons of the outer planets were relatively uninteresting and geologically dead. They assumed that (planetary) heat sources were not sufficient to have melted the mantles of these moons enough to provide a source of liquid—or even semi-liquid—ice or silicate slurries.

MARS EXPLORATION ROVER MISSION

In the summer of 2003, NASA launched identical twin Mars rovers that were to operate on the surface of the Red Planet during 2004. *Spirit* (Mars Exploration Rover [MER]-A) was launched by a Delta II rocket from Cape Canaveral on June 10, 2003, and successfully landed on Mars on January 4, 2004. *Opportunity* (MER-B) was launched from Cape Canaveral on July 7, 2003, by a Delta II rocket and successfully landed on the surface of Mars on January 25, 2004. Both landings used the successful airbag bounce-and-roll arrival demonstrated during the Mars Pathfinder mission.

Following arrival on the surface of the Red Planet, each rover drove off and began its surface exploration mission in a decidedly different location. Communications back to Earth has been accomplished primarily with Mars-orbiting spacecraft, such as the *Mars Odyssey 2001,* serving as data relays.

Spirit landed in Gusev Crater, which is roughly 15 degrees south of the Martian equator. NASA mission planners selected Gusev Crater because it had the appearance of a crater lakebed. *Opportunity* landed at Terra Meridiani—a region of Mars that is also known as the Hematite Site because this location

This interesting mosaic image was taken by the navigation camera on NASA's Mars Exploration Rover *Spirit* on January 4, 2004. NASA scientists have reprocessed the image to project a clear overhead view of the robot rover and its lander mother spacecraft on the surface of Mars. *(NASA/JPL)*

displayed evidence of coarse-grained hematite, an iron-rich mineral that typically forms in water. Among this mission's principal scientific goals is the search for and characterization of a wide range of rocks and soils that hold clues to past water activity on Mars. By the end of 2006, both rovers continued to function and move across Mars far beyond the primary mission goal of 90 days.

Each rover has a mass of 407 pounds (185 kg) and has successfully traveled up to 330 feet (100 m) per sol (Martian day) across the surface of the planet. *Spirit* and *Opportunity* each carry an identical complement of sophisticated instruments that allows them to search for evidence that liquid water was present on the surface of Mars in ancient times. As part of NASA's overall exploration strategy, the rovers landed in different regions of Mars. Immediately after landing, each rover performed reconnaissance of the particular landing site by taking panoramic (360-degree) visible (color) and infrared images. Then, using images and spectra taken daily by the rovers, NASA scientists at the Jet Propulsion Laboratory used telecommunications and teleoperations to supervise the overall scientific program. With intermittent human guidance, the pair of mechanical explorers functioned like robot prospectors—examining particular rocks and soil targets and evaluating composition and texture at the microscopic level.

✧ Exploring Small Bodies in the Solar System

Just two decades ago, scientists did not have a lot of specific information about the small bodies—such as comets and asteroids—in the solar system. There was a great deal of speculation about the true nature of a comet's nucleus, and no one had ever seen the surface of an asteroid up close. All that changed very quickly when robot spacecraft missions flew past, imaged, sampled, probed, and even landed on several of these interesting celestial objects. This section of the chapter discusses some of the most significant small-body missions that have taken place.

Asteroids and comets are believed to be the ancient remnants of the earliest years of the formation of the solar system, which took place more than 4 billion years ago. From the beginning of life on Earth to the spectacular collision of Comet Shoemaker-Levy 9 with Jupiter (in July 1994), these so-called small bodies influence many of the fundamental processes that have shaped the planetary neighborhood in which Earth resides.

Scientists currently believe that asteroids are the primordial material that was prevented by Jupiter's strong gravity from accreting (accumulating) into a planet-size body when the solar system was born about 4.6 billion years ago. It is estimated that the total mass of all the asteroids (if assembled together) would comprise a celestial body about 932 miles (1,500 km) in diameter—an object less than half the size (diameter) of the Moon.

NASA's *Galileo* spacecraft was the first to observe an asteroid close-up, flying past main-belt asteroids Gaspra and Ida in 1991 and 1993, respectively. Gaspra and Ida proved to be irregularly shaped objects, rather like potatoes, riddled with craters and fractures. The *Galileo* spacecraft also discovered that Ida had its own moon—a tiny body called Dactyl in orbit around its parent asteroid. Astronomers suggest Dactyl may be a fragment from past collisions in the asteroid belt.

A comet, as confirmed by space missions, is a dirty ice-rock consisting of dust, frozen water, and gases that orbits the Sun. As a comet approaches

An artist's rendering of the *Near-Earth Asteroid Rendezvous (NEAR)* spacecraft's encounter with the asteroid Eros, which began on February 14, 2000. After going into orbit around Eros and examining the asteroid for a year, the mission ended when the *NEAR-Shoemaker* spacecraft touched down in the saddle region of the minor planet on February 12, 2001. NASA renamed this robot spacecraft *NEAR-Shoemaker* in honor of the American geologist and astronomer Eugene M. Shoemaker (1928–97). *(NASA/Johnson Space Center; artist, Pat Rawlings)*

the inner solar system from deep space, solar radiation causes its frozen materials to vaporize (sublime), creating a coma and a long tail of dust and ions. While the accompanying coma and tail may be very large, comet nuclei generally have diameters of only a few tens of miles or less. Scientists think these icy planetesimals are the remainders of the primordial material from which the outer planets were formed billions of years ago.

NEAR-EARTH ASTEROID RENDEZVOUS MISSION

NASA's *Near-Earth Asteroid Rendezvous* (*NEAR*) spacecraft was the first scientific mission dedicated to the exploration of an asteroid. It was launched on February 17, 1996, from Cape Canaveral by a Delta II expendable launch vehicle. The *NEAR* spacecraft was equipped with an X-ray/gamma-ray spectrometer, a near-infrared imaging spectrograph, a multispectral camera fitted with a charge-coupled device (CCD) imaging detector, a laser altimeter, and a magnetometer. The primary goal of this mission was to rendezvous with and achieve orbit around the near-Earth asteroid Eros (also called 433 Eros).

Eros is an irregularly shaped S-class asteroid about 8 × 8 × 20.5 miles (13 × 13 × 33 km) in size. This asteroid, the first near-Earth asteroid to be found, was discovered on August 13, 1898, by the German astronomer Gustav Witt (1866–1946). In Greek mythology, Eros (Roman name: Cupid) was the son of Hermes (Roman name: Mercury) and Aphrodite (Roman name: Venus) and served as the god of love.

After launch and departure from Earth's orbit, *NEAR* entered the first part of its cruise phase. The robot spacecraft spent most of this phase in a minimal activity (hibernation) state that ended a few days before the successful flyby of the asteroid Mathilde on June 27, 1997. During that encounter, the spacecraft flew within 745 miles (1,200 km) of Mathilde at a relative velocity of 6.2 miles (9.93 km) per second. Imagery and other scientific data were collected.

The original mission plan was to rendezvous with and achieve orbit around Eros in January 1999 and then to study the asteroid for approximately one year. However, a software problem caused an abort of the first-encounter rocket-engine burn, and NASA revised the mission plan to include a flyby of Eros on December 23, 1998. This flyby was then followed by an encounter and orbit in February 2000.

The *NEAR-Shoemaker* spacecraft caught up with the asteroid Eros on February 14, 2000, and orbited the small body for a year, studying its surface, orbit, mass, composition, and magnetic field. (NASA renamed the spacecraft *NEAR-Shoemaker* in honor of American astronomer and geologist Eugene M. Shoemaker [1928–97], following his untimely death in an automobile accident on July 18, 1997.) Then, in February 2001, mission controllers guided the spacecraft to the first-ever landing on an asteroid.

The mission ended with a touchdown in the saddle region of Eros on February 12, 2001.

GIOTTO MISSION

The European Space Agency's scientific spacecraft, called *Giotto,* was launched on July 2, 1985, from the agency's Kourou, French Guiana, launch site by an Ariane 1 rocket. On March 13, 1986, the spacecraft successfully encountered Comet Halley at a closest approach of 370 miles (596 km) distance. This historic encounter took place during the famous comet's predicted 1986 return to the inner solar system. After the successful encounter, *Giotto* went into hibernation and awakened in 1992 to study the comet Grigg-Skjellerup as part of an extended science mission.

Giotto's scientific payload consisted of 10 hardware experiments: a narrow-angle camera; three mass spectrometers for neutral particles, ions, and dust particles; various dust detectors; a photopolarimeter; and a set of plasma experiments. During the Comet Halley encounter, all experiments performed well and produced an enormous number of significant scientific results. Perhaps the most important accomplishment was the clear identification of the comet's nucleus and the confirmation of the hypothesized dirty-snowball (that is, rock and ice) model.

The European Space Agency named the *Giotto* spacecraft after the Italian painter Giotto di Bondone (1266–1337), who apparently witnessed the 1301 passage of Comet Halley. The Renaissance artist included the first scientific representation of this famous comet in his renowned fresco *Adoration of the Magi,* which can be found in the Scrovegni Chapel in Padua, Italy.

STARDUST MISSION

The primary objective of the Stardust mission was to fly by Comet Wild 2 and collect samples of dust and volatiles in the coma of this comet. NASA launched the *Stardust* spacecraft from Cape Canaveral Air Force Station, Florida, on February 7, 1999—using an expendable Delta II rocket. Following launch, the spacecraft successfully achieved an elliptical, heliocentric orbit. By mid-summer 2003, it had completed its second orbit of the Sun. The spacecraft then successfully flew by the nucleus of Comet Wild 2 on January 2, 2004.

When *Stardust* flew past the comet's nucleus it did so at an approximate relative velocity of 3.8 miles (6.1 km) per second. At closest approach during the encounter, the spacecraft came within 155 miles (250 km) of the comet's nucleus and returned images of the nucleus. The spacecraft's dust-monitor data indicate that many particle samples were collected. Based on telemetry data from the spacecraft, mission scientists estimated that during the close encounter, *Stardust* captured thousands of particles and volatiles of cometary material.

Stardust next traveled on a trajectory that brought it near Earth in January 2006. The comet-material samples that were collected were stowed and sealed in the special sample-storage vault of the reentry capsule carried onboard the *Stardust* spacecraft. As the robot spacecraft flew past Earth in mid-January 2006, it successfully ejected the sample capsule. The sample capsule descended through Earth's atmosphere on January 15, 2006, and was successfully recovered after soft landing in the Utah desert. Scientists are now examining the contents of the return capsule. These tiny pieces of extraterrestrial material are the first pristine samples of cometary coma materials and (suspected) interstellar materials, that were captured in space by a robot spacecraft and returned to Earth for scientific study.

DEEP IMPACT MISSION

In early July 2005, NASA's *Deep Impact* robot spacecraft performed a complex experiment in space that probed beneath the surface of a comet and helped reveal some of the secrets of its interior. As a larger flyby spacecraft released a smaller impactor spacecraft into the path of Comet Tempel 1, the experiment became one of a cometary bullet chasing down a spacecraft bullet (the penetrator), while a third spacecraft bullet (the flyby robot) sped along to watch.

The greatest challenge for the engineers who created the *Deep Impact* flight system and its collection of science instruments was to target and successfully hit the 3.7-mile- (6-km-) diameter nucleus of Comet Tempel 1. Traveling at a relative velocity of 6.2 miles (10 km) per second and released by the flyby spacecraft from a distance of about 537,000 miles (864,000 km), the self-guided impactor had to strike in an area on the sunlit side of the nucleus to allow the flyby spacecraft's science instruments to take images of the collision and its aftermath.

The *Deep Impact* flight system consisted of two robot spacecraft: the flyby spacecraft and the impactor. Each spacecraft had its own instruments and capabilities to receive and transmit data. The flyby spacecraft carried the primary imaging instruments and hauled the impactor to the vicinity of the comet's nucleus. Serving as a mother spacecraft for the mission, the *Deep Impact* flyby spacecraft released the impactor about 24 hours prior to comet impact. The flyby spacecraft then received data from the impactor as it traveled to the target, used its onboard instruments to record images of the impactor-comet collision, observed post-collision phenomena (including the resultant crater and ejected materials), and then transmitted all the science data back to Earth.

This unusual and interesting mission began with a successful launch from Cape Canaveral on January 12, 2005. Following launch, the *Deep*

Impact flyby spacecraft and the hitchhiking (cojoined) impactor transferred into a heliocentric orbit and rendezvoused with Comet Tempel 1 in early July. On July 3, the flyby spacecraft released the impactor and executed a velocity decrease and realignment maneuver to better observe

This is an artist's rendering of the nucleus of Comet Tempel 1, which is depicted in a simulated infrared view—based on information provided by the *Spitzer Space Telescope* (SST). Data from the *Spitzer Space Telescope* indicate that the comet is a matte black object roughly 8.7 × 2.5 miles (14 × 4 km) in size. *(NASA/JPL-Caltech)*

the impact. On July 4 (about 24 hours later), the impactor smashed into the comet's nucleus.

As planned, *Deep Impact's* impactor successfully collided with Comet Tempel 1 on the sunlit side. A camera on the impactor spacecraft captured and relayed images of the comet's nucleus until just seconds before the collision. Upon impact there was a brilliant and rapid release of dust that momentarily saturated the cameras onboard the *Deep Impact* flyby spacecraft. Audiences around the world watched this spectacular event, as dramatic images were shown in near–real time on NASA TV and broadcast over the Internet. All available NASA-orbiting telescopes, including the *Hubble Space Telescope,* the *Chandra X-ray Observatory,* and the *Spitzer Space Telescope,* observed the unique event that occurs when Earth strikes a comet.

The flyby spacecraft was approximately 6,200 miles (10,000 km) away at the time of impact and began collecting images about 60 seconds before impact. At approximately 600 seconds after impact, the flyby spacecraft was about 2,500 miles (4,000 km) from the nucleus, and observations of the crater started and continued until closest approach

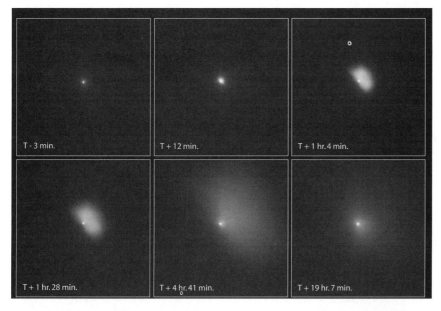

This series of *Hubble Space Telescope* images captures the ejection of a bright plume of dust following the July 4, 2005, collision between an 820-pound (373-kg) projectile released by NASA's *Deep Impact* spacecraft and Comet Tempel 1. The image sequence dramatically shows the evolution of material that was blasted off the comet as it expands and diffuses into space. *(NASA, ESA, P. Feldman [Johns Hopkins University], and H. Weaver [Johns Hopkins University Applied Physics Lab])*

to the nucleus at a distance of about 310 miles (500 km). Sixteen minutes after impact, imaging ended, as the flyby spacecraft aligned itself to cross the inner coma. Within 21 minutes, the crossing of the inner coma was complete, and the *Deep Impact* flyby spacecraft once again aligned itself—this time to look back at the comet. After 50 minutes, the flyby spacecraft began to play back all the accumulated scientific data to scientists on Earth. The *Deep Impact* flyby spacecraft is now in a hibernation (or sleep) mode, awaiting a possible wake-up call for further scientific investigations. The impact, while powerful, was not forceful enough to make an appreciable change in the comet's orbital path around the Sun. Ice, heated by the energy of the impact, vaporized, and dust and debris were ejected from the crater. As scientists sort through the data gathered by the flyby spacecraft, they will to learn a little more about the structure of a comet's interior and whether it is different substantially different from its surface.

✦ New Horizons Pluto–Kuiper Belt Flyby Mission

Originally conceived as the *Pluto Fast Flyby,* NASA's *New Horizons Pluto–Kuiper belt Flyby* spacecraft was successfully launched from Cape Canaveral on January 19, 2006, and is now traveling on its way to Pluto. This reconnaissance-type exploration mission will help scientists understand the interesting, yet poorly understood, icy worlds at the edge of the solar system. If all goes well over the next nine years, the first spacecraft flyby of Pluto and Charon—the frigid double-planet system—will take place sometime in the summer of 2015. The mission will then continue beyond Pluto and visit one or more Kuiper belt objects (of opportunity) by 2026. The spacecraft's long journey will help resolve some basic questions about the surface features and properties of these icy bodies as well as their geology, interior makeup, and atmospheres.

With respect to the Pluto-Charon system, some of the major scientific objectives include the characterization of the global geology and geomorphology of Pluto and Charon, the mapping of the composition of Pluto's surface, and the determination of the composition and structure of Pluto's transitory atmosphere. It is intended that the spacecraft will reach Pluto before the tenuous Plutonian atmosphere can refreeze onto the surface as the planet recedes from its 1989 perihelion. Studies of the double-planet system will actually begin some 12 to 18 months before the spacecraft's closest approach to Pluto in mid-2015. The modestly sized spacecraft has no deployable structures and receives all its electric power

An artist's concept of NASA's *New Horizons* spacecraft during its planned encounter with the dwarf planet Pluto (foreground) and its relatively large moon Charon in the summer of 2015. (Astronomical observations in 2005 suggest that Pluto may also have two smaller moons, which do not appear in this rendering.) A long-lived, plutonium-238 fueled, radioisotope-thermoelectric generator (RTG) system (cylinder on lower-left portion of the spacecraft) provides electric power to the robot spacecraft as it flies past these distant icy worlds billions of miles from the Sun. As depicted here, one of the spacecraft's most prominent features is a 6.9-foot- (2.1-m-) diameter disk antenna through which *New Horizons* can communicate with scientists on Earth from as far as 4.8 billion miles (7.5 billion km) away. Following its encounter with Pluto, the *New Horizons* spacecraft hopes to explore one or several icy planetoid targets of opportunity in the Kuiper belt. *(NASA/JPL)*

from long-lived radioisotope thermoelectric generators that are similar in design to those used on the *Cassini* spacecraft now orbiting Saturn.

This important mission will complete the initial scientific reconnaissance of the solar system with robot spacecraft. At present Pluto is the most poorly understood major planet in the solar system. As some scientists speculate, the tiny planet may even be regarded as the largest member of the family of primitive icy objects that reside in the Kuiper belt. In addition to the first close-up view of Pluto's surface and atmosphere, the spacecraft will obtain gross physical and chemical surface properties of Pluto,

An artist's rendering of the Pluto system as seen from the surface of one of the new moons. In 2005, using NASA's *Hubble Space Telescope* to study the icy dwarf planet, astronomers discovered that Pluto has three moons: Charon, Nix, and Hydra. *(NASA)*

Charon, and (possibly) several Kuiper belt objects. A successful launch was absolutely necessary between mid-January and early February 2006, if the robot spacecraft was to take advantage of a celestial mechanics opportunity to receive a gravity assist from Jupiter in 2007. This gravity-assist maneuver at Jupiter will increase the spacecraft's speed and take years off a very long interplanetary journey to the end of the solar system.

To add a little more scientific excitement to an already incredible planetary mission, investigators, using NASA's *Hubble Space Telescope* to observe Pluto on May 15, 2005, discovered that the icy planet has not one, but three moons.

The two new Plutonian moons, designated Hydra (S/2005 P1) and Nix (S/2005 P2), are approximately 27,000 miles (44,000 km) away from Pluto—in other words, two or three times as far from Pluto as Charon.

These moons are tiny. Their diameters are estimated to be between 40 and 125 miles (64 and 200 km). For comparison, Charon is about 730 miles (1,170 km) wide, while Pluto itself has a diameter of about 1,410 miles (2,270 km).

The second interesting event concerning Pluto took place on August 24, 2006, when the members of the International Astronomical Union (IAU) met in Prague, the Czech Republic, and decided (by vote) to demote Pluto from its traditional status as one of the nine major planets and place the object into a new planetary class, called a *dwarf planet*. The IAU decision now leaves the solar system with eight major planets and three dwarf planets: Pluto (which serves as the prototype dwarf planet), Ceres (the largest asteroid), and the large, distant Kuiper belt object identified as 2003 UB313 (nicknamed "Xena"). Astronomers anticipate the discovery of other dwarf planets in the distant parts of the solar system.

The IAU decision to downsize Pluto in astronomical status from a major planet to a dwarf planet does not affect the scientific objectives of NASA's New Horizons mission. In early September 2006, NASA engineers tested the spacecraft's long-range reconnaissance imager (LORRI) by collecting an image of the Messier 7 star cluster. The electronic snapshot also meant that all seven science instruments on the *New Horizons* spacecraft have now operated properly in space and returned good data since the spacecraft was launched. As of September 2006, the robot spacecraft remains "go" for its historic July 2015 flyby encounter with the now "dwarf" planet Pluto.

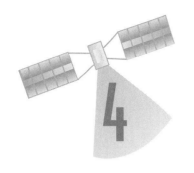

Optical Astronomy and the *Hubble Space Telescope*

4

Optical astronomy is the branch of astronomy that uses radiation in the visible-light portion of the electromagnetic spectrum—nominally 4,000 to 7,000 Å (400 to 700 nm) in wavelength—to study celestial objects and cosmic phenomena. Astronomers use visual radiation collected by telescopes on the ground and in space to study planets, stars and stellar evolution, normal and active galaxies, and the large-scale structure of the universe.

Naked-eye astronomy, the practice of astronomy without the assistance of an optical telescope, is inherently optical astronomy because the human eye is sensitive to visible light, and it is information carried by this light that helps human beings form a visual awareness of the universe and its contents. All astronomical observations before the early 17th century were naked-eye studies of the heavens. However, naked-eye astronomers often used ancient astronomical instruments, such as the astrolabe, to locate objects and track celestial objects. The Danish astronomer Tycho Brahe (1546–1601) is considered the last and greatest naked-eye astronomer.

Telescopic astronomy is also part of optical astronomy, if the instrument is sensitive to sources of visible light. Of course many modern optical telescopes, especially space-based systems, such as the *Hubble Space Telescope,* and large-diameter ground-based facilities (which are located on tall mountaintops) have capabilities that extend beyond the narrow visible portion of the electromagnetic spectrum into the near-infrared region on one side and into the ultraviolet region on the other side of the visible spectrum. Above about 2.0 micrometers wavelength (which is in the infrared region), increasing thermal emissions from the optical telescope itself and (in the case of ground-based telescopes) from Earth's atmosphere, require the use of special detectors and "chilled" telescope designs. So, astronomers practice infrared astronomy (discussed in chapter 7) using specially designed space-based telescopes or ground-based telescopes, which are usually perched on a high mountaintop above most of Earth's meddlesome atmosphere.

A variety of other related disciplines complement the traditional detection and measurement areas of optical astronomy. Three of these companion disciplines are astrophotography, optical interferometry, and spectroscopy. Astrophotography involves the use of photographic techniques to create images of celestial bodies. Astronomers are now replacing light-sensitive photographic emulsions with charged-coupled devices to create digital images in the visible, infrared, and ultraviolet portions of the electromagnetic spectrum. The application of photography in observational astronomy began in the middle of the 19th century and has exerted great influence on the field ever since.

Optical interferometry is based on the use of an instrument called an interferometer. In the early 1880s, with financial support from the Scottish-American inventor Alexander Graham Bell (1847–1922), the American Nobel laureate Albert Abraham Michelson (1852–1931) constructed a precision optical interferometer (sometimes called the Michelson interferometer). His instrument split a beam of light into two paths and then reunited the two separate beams. Should either beam experience travel across different distances or at different velocities (due to passage through different media), the reunited beam would appear out of phase and produce a distinctive arrangement of dark and light bands, called an interference pattern.

In principle, therefore, the optical interferometer produces and measures interference fringes from two or more coherent wave trains from the same optical radiation (light) source. Astronomers achieve high angular-resolution measurements by combining the light signals received from a source by at least two widely separated (optical) telescopes. Optical interferometry is used to precisely measure wavelengths, to measure the angular width of sources, and to determine the angular position of sources.

In 1859 the German chemist Robert Wilhelm Bunsen (1811–99) collaborated with the German physicist Gustav Robert Kirchhoff (1824–87) in the development of spectroscopy. Their pioneering work revolutionized astronomy by allowing scientists to determine the chemical composition of distant celestial bodies. Spectroscopy involves the study of spectral lines from different atoms and molecules. Modern astronomers use emission spectroscopy to infer the material composition of the objects that emitted the light and absorption spectroscopy to infer the composition of the intervening medium.

Astronomers traditionally have used the term *observable universe* to describe the portion of the entire universe that can be seen from Earth.

Originally—from 1610 (Galileo Galilei's first telescopic observations of the heavens) up until the early part of the 20th century—this term was restricted to the portion of the universe that could be detected and studied by the visible light emitted by celestial objects (such as stars) or reflected from celestial objects (such as moons and planets). Starting in the middle of the 20th century, astronomers expanded the term to mean that portion of universe which is detectable across all portions of the electromagnetic spectrum (including gamma-ray, X-ray, ultraviolet, visible-light, infrared, microwave, and radio-frequency), as well as through the study of energetic cosmic rays and neutrinos. Logically excluded from this current meaning of the term *observable universe* is the postulated dark matter (or missing mass) of the universe.

ANDERS ÅNGSTRÖM

Anders Ångström was a Swedish physicist and solar astronomer who performed pioneering spectral studies of the Sun. In 1862 he discovered that hydrogen was present in the solar atmosphere and went on to publish a detailed map of the Sun's spectrum, covering other elements present as well. A special unit of wavelength, the angstrom (symbol: Å), now honors his accomplishments in spectroscopy and astronomy.

Ångstrom was born in Lögdö, Sweden, on August 13, 1814. He studied physics and astronomy and graduated in 1839 with his doctorate from the University of Uppsala. This university was founded in 1477 and is the oldest of the Scandinavian universities. Upon graduation Ångström joined the university faculty as a lecturer in physics and astronomy. For more than three decades, he remained at this institution, serving it in a variety of academic and research positions. For example, in 1843 he became an astronomical observer at the famous Uppsala Observatory—the observatory founded in 1741 by Anders Celsius (1701–44). In 1858 Ångström became chairperson of the physics department and remained a

professor in that department for the remainder of his life.

He performed important research in heat transfer, spectroscopy, and solar astronomy. With respect to his contributions in heat-transfer phenomena, Ångstrom developed a method to measure thermal conductivity by showing that it was proportional to electrical conductivity. He was also was one of the 19th century's pioneers of spectroscopy. Ångstrom observed that an electrical spark produces two superimposed spectra. One spectrum is associated with the metal of the electrode generating the spark, while the other spectrum originates from the gas through which the spark passes.

Ångstrom applied Leonhard Euler's (1707–83) resonance theorem to his experimentally derived atomic spectra data and discovered an important principle of spectral analysis. In his paper "Optiska Undersökningar" (Optical investigations), which he presented to the Swedish Academy in 1853, Ångstrom reported that an incandescent (hot) gas emits light at precisely the same wavelength as it absorbs light when it is cooled. This finding represents Ång-

✧ Stars and Their Life Cycles

A star is essentially a self-luminous ball of very hot gas that generates energy through thermonuclear fusion reactions that take place in its core. Astronomers classify stars as either normal or abnormal. Normal stars, such as the Sun, shine steadily. These stars exhibit a variety of colors: red, orange, yellow, blue, and white. Most stars are smaller than the Sun, and many stars even resemble it. However, a few stars are much larger than the Sun. In addition astronomers have observed several types of abnormal stars, including giants, dwarfs, and a variety of variable stars.

Most stars can be put into one of several general spectral types called O, B, A, F, G, K, and M. In the mid- to late 19th century,

strom's finest research work in spectroscopy, and his results anticipated the spectroscopic discoveries of Gustav Kirchhoff that led to the subsequent formulation of Kirchhoff's laws of radiation. Ångstrom was also able to demonstrate the composite nature of the visible spectra of various metal alloys.

His laboratory activities at the University of Uppsala also provided Ångstrom with hands-on experience in the emerging field of spectroscopy necessary to accomplish his pioneering observational work in solar astronomy. By 1862 Ångstrom's initial spectroscopic investigations of the solar spectrum enabled him to announce his discovery that the Sun's atmosphere contained hydrogen. In 1868 he published *Recherches sur le spectre solaire* (Researches on the solar spectrum), his famous atlas of the solar spectrum, containing his careful measurements of approximately 1,000 Fraunhofer lines. Unlike other pioneering spectroscopists such as Robert Bunsen and Gustav Kirchhoff who used an arbitrary measure, Ångstrom precisely measured the corresponding wavelengths in units equal to one 10-millionth of a meter.

Ångstrom's map of the solar spectrum served as a standard of reference for astronomers for

nearly two decades. In 1905 the international scientific community honored his contributions by naming the unit of wavelength he used the angstrom —one angstrom corresponds to a length of 10^{-10} meter. Physicists, spectroscopists, and microscopists use the angstrom when they discuss the visible-light portion of the electromagnetic spectrum. The human eye is sensitive to electromagnetic radiation with wavelengths between 400 and 700 nanometers. These numbers are very small, so scientists often find it more convenient to use the angstrom (unit) in their technical discussions. For example, the range of human vision can be expressed as ranging between 4,000 and 7,000 angstroms.

In 1867 Ångstrom became the first scientist to examine the spectrum of the aurora borealis (northern lights). Because of this pioneering work, his name is sometimes associated with the aurora's characteristic bright yellow–green light. He was a member of the Royal Swedish Academy (Stockholm) and the Royal Academy of Sciences of Uppsala. In 1870 Ångstrom was elected a Fellow of the Royal Society in London, from which society he received the prestigious Rumford Medal in 1872. He died in Uppsala on June 21, 1874.

astronomers established this classification sequence by placing observable stars in order of their decreasing surface temperature. O stars are very hot, large blue stars with surface temperatures ranging from 28,000 K to 40,000 K and more. Sometimes called ultraviolet stars, O-type stars are the hottest observable stars and have very short lifetimes, typically between 3 and 6 million years. B stars are large, hot blue stars with surfaces temperatures ranging between 11,000 K and 28,000 K. The star Rigel is an example. A-type stars are white to blue-white in color. The surface temperatures of A stars range from 7,500 K to 11,000 K. Vega, Sirius, and Altair are A stars. F stars are white stars with surface temperatures ranging from 6,000 K to 7,500 K. The stars Canopis and Polaris are examples. G stars are yellow stars with surface temperatures ranging from 5,000 K to 6,000 K. The solar system's parent star, the Sun, is an example. K stars are orange-red stars with surface temperatures ranging from 3,500 K to 5,000 K. Arcturus and Aldebaran are examples. Finally, the coolest stars are M stars. These stars appear red and have surface temperatures of less than 3,500 K. Antares and Betelgeuse are examples.

In the mid-1940s, astronomers found it useful to further subdivide

Astronomers call this swarm of stars M80 (or NGC 6093). M80 is one of the densest of the almost 150 known star clusters in the Milky Way Galaxy. Located abut 28,000 light-years from Earth, M80 contains hundreds of thousands of stars, all held together in a globular cluster by their mutual gravitational attraction. Astronomers find the study of such globular clusters especially useful in their investigation of the life cycle of stars, since all the stars in a cluster have the same age (about 13.7 to 14 billion years) but cover a range of stellar masses. Every star visible in this image (collected by the *Hubble Space Telescope* in 1999) is either more highly evolved than or, in a few rare cases, more massive than the Sun. *(NASA/STScI)* each lettered spectral classification into 10 subdivisions, denoted by the numbers zero to nine. The convention used within modern astronomy is that the hotter the star, the lower the number in this subdivision. As a result, astronomers classify the Sun as a G2 star. This means the Sun is a bit hotter than a G3 star and a bit cooler than a G1 star. Betelgeuse is an M2 star, and Vega is an A0 star.

BETELGEUSE

Betelgeuse is the red supergiant star called Alpha Orinois (α Ori) by astronomers. This prominent star marks the right shoulder in the constellation Orion of the great hunter from Greek mythology. Betelgeuse has a spectral classification of M2Ia and is a semi-regular variable with a period of about 5.8 years. It has a normal apparent magnitude (visual) range between 0.3 and 0.9 (about +0.41 average) but can vary between 0.15 and 1.3. This huge star (about 500 times the diameter of the Sun) is approximately 490 light-years away from Earth. The two brightest stars in the constellation Orion were designated α Ori (Betelgeuse) and β Ori (Rigel) in order of their apparent brightness, using the notation introduced by German astronomer Johann Bayer 1572–1625) in the 17th century. However, more

Size of Star

Size of Earth's Orbit

Size of Jupiter's Orbit

This is the first direct image of a star other than the Sun. Called Alpha Orionis, or Betelgeuse, the star is a red supergiant—a Sun-like star near the end of its life. The *Hubble Space Telescope* image (collected on January 15, 1996) reveals a huge ultraviolet atmosphere with a mysterious hot spot on the stellar behemoth's surface. The enormous bright spot, more than 10 times the diameter of Earth, is at least 2,000 K hotter than the rest of star's surface. *(Andrea Dupree, Harvard-Smithsonian CfA; Ronald Gilliland, STScI; NASA; and ESA)*

precise recent observations indicate that Rigel is actually brighter than Betelgeuse. Nevertheless, to avoid confusion and preserve astronomical tradition, astronomers still retain the previous stellar nomenclature.

Modern astronomers have also found it helpful to categorize stars by luminosity. This classification scheme is based upon a single spectral property—the width of a star's spectral-lines. From astrophysics, scientists know that spectral-line width is sensitive to the density conditions in a star's photosphere. In turn, astronomers can correlate a star's atmospheric density to its luminosity. The use of luminosity classes allows astronomers to distinguish giants from main-sequence stars, supergiants from giants, and so forth. The stellar luminosity classes used by astronomers are as follows: class Ia stars are bright supergiants; class Ib are supergiants; class II are bright giants; class III are giants; class IV are subgiants; and class V are main-sequence stars and dwarfs.

The Sun is approximately 0.87 million miles (1.4 million km) in diameter and has an effective surface temperature of about 5,800 K. (Chapter 9 provides a more detailed discussion about the Sun.) The Sun, like other stars, is a giant nuclear furnace in which the temperature, pressure, and density are sufficient to cause light nuclei to join together, or fuse. For example, deep inside the solar interior, hydrogen, which makes up 90 percent of the Sun's mass, is fused into helium atoms, releasing large amounts of energy that eventually works its way to the surface and then is radiated throughout the solar system. The Sun is currently in a state of balance, or equilibrium, between two competing forces: gravity (which wants to pull all its mass inward) and the radiation pressure and hot gas pressure resulting from the thermonuclear reactions (which push outward).

Many stars in the galaxy appear to have companions, with which they are gravitationally bound in binary, triple, or even larger systems. Compared to other stars throughout the Milky Way Galaxy, the Sun is slightly unusual. It does not have a known stellar companion. However, in an attempt to explain an apparent cosmic catastrophe cycle that occurred on Earth about 65 million years ago, some astrophysicists have postulated the existence of a very distant, massive, dark companion called Nemesis.

Through careful telescopic observations and spectroscopic measurements, astronomers and astrophysicists have discovered what appears to be the life cycle of stars. Stars originate by the condensation of enormous clouds of cosmic dust and hydrogen gas, called nebulae. Gravity is the dominant force behind the birth of a star. According to Newton's universal law of gravitation, all bodies attract each other in proportion to their masses and distance apart. The dust and gas particles found in these huge interstellar clouds attract each other and gradually draw closer together. Eventually, enough of these particles join together to form a central clump that is sufficiently massive to bind all the other parts of the cloud by gravitation. At this point, the edges of the cloud start to collapse inward, separating it from the remaining dust and gas in the region.

Initially, the cloud contracts rapidly because the thermal energy release related to contraction is radiated outward easily. However, when the cloud grows smaller and denser, the heat released at the center cannot escape to the outer surface immediately. This causes a rapid rise in internal temperature, slowing down but not stopping the relentless gravitational contraction.

The actual birth of a star occurs when its interior becomes so dense and its temperature so high that thermonuclear fusion occurs. The heat released in thermonuclear fusion reactions is greater than that released through gravitational contraction, and fusion becomes the star's primary energy-producing mechanism. Gases heated by nuclear fusion at the cloud's center begin to rise, counterbalancing the inward pull of gravity on the outer layers. The star stops collapsing and reaches a state of equilibrium between outward and inward forces. At this point, the star has become what astronomers and astrophysicists call a main-sequence star. Like the Sun, it will remain in this state of equilibrium for billions of years, until all the hydrogen fuel in its core has been converted into helium.

How long a star remains on the main sequence, burning hydrogen for its fuel, depends mostly on its mass. The Sun has an estimated main-sequence lifetime of about 10 billion years, of which approximately 5 billion years have now passed. Larger stars burn their fuels faster and at much higher temperatures. These stars, therefore, have short main-sequence lifetimes, sometimes as little as 1 million years. In comparison, the red dwarf stars, which typically have less than $\frac{1}{10}$ the mass of the Sun, burn up so slowly that trillions of years must elapse before their hydrogen supply is exhausted. When a star has used up its hydrogen fuel, it leaves the normal state or departs the main sequence. This happens when the core of the star has been converted from hydrogen to helium by thermonuclear reactions that have taken place.

When the hydrogen fuel in the core of a main-sequence star has been consumed, the core starts to

This spectacular image (taken by NASA's *Hubble Space Telescope* in 1995) shows a "star birth" cloud in M16—an open cluster in the constellation Serpens (Ser). Such dense clouds of gas—called giant molecular clouds (GMCs)—are located primarily in the spiral arms of galaxies. Astronomers believe the GMC is the birthplace of stars. A new star is born when gravitational attraction causes a denser region in the cloud to form into a protostar. The protostar continues to contract until its compressed central core initiates hydrogen and helium burning. At that point, it becomes a main–sequence star. *(NASA)*

collapse. At the same time, the hydrogen fusion process moves outward from the core into the surrounding outer regions. There the process of converting hydrogen into helium continues, releasing radiant energy. But as this burning process moves into the outer regions, the star's atmosphere expands greatly, and it becomes a red giant. The term *giant* is quite appropriate. As an imaginary or thought experiment, if scientists put a red giant where the Sun is now, the innermost planet, Mercury, would be engulfed by it; similarly, if they put a larger red supergiant there, the supergiant star would extend past the orbit of Mars.

As the star's nuclear evolution continues, it might become a variable star, pulsating in size and brightness over periods of several months to years. The visual brightness of such an abnormal star might now change by a factor of 100, while its total energy output varies by only a factor of 2 or 3.

As an abnormal star grows, its contracting core may become so hot that it ignites and burns nuclear fuels other than hydrogen, beginning with the helium created in millions to perhaps billions of years of main-sequence burning. The subsequent behavior of such a star is complex, but in general it can be characterized as a continuing series of gravitational contractions and new nuclear-reaction ignitions. Each new series of fusion reactions produces a succession of heavier elements in addition to releasing large quantities of energy. For example, the burning of helium produces carbon, the burning of carbon produces oxygen, and so forth.

Finally, when nuclear burning no longer releases enough radiant energy to support the giant star, it collapses and its dense central core becomes either a compact white dwarf or a tiny neutron star. This collapse also may trigger an explosion of the star's outer layers—displaying itself as a supernova. In exceptional cases with very massive stars, the core (or perhaps even the entire star) might become a black hole.

When a star like the Sun has burned all the nuclear fuels available, it collapses under its own gravity until the collective resistance of the electrons within it finally stops the contraction process. The dead star has become a white dwarf and may now be about the size of Earth. Its atoms are packed so tightly together that a fragment the size of a sugar cube would have a mass of thousands of kilograms. The white dwarf then cools for perhaps several billion years, going from white, to yellow, to red, and finally becomes a cold, dark sphere sometimes called a black dwarf. (Note that the white dwarf does not experience thermonuclear burning; rather, its light comes from a very hot, thin gaseous atmosphere that gradually dissipates its heat to space.) Astrophysicists estimate that there are over 10 billion white dwarf stars in the Milky Way Galaxy alone, many of which will eventually become black dwarfs. A black dwarf represents the cold remains of a white dwarf star that no longer emits visible radiation. This object can also be a nonradiating ball of interstellar gas that has contracted under gravitation but contains too little mass to initiate nuclear fusion. The fate of becoming first a white dwarf, then millions of years later, a cold black dwarf, appears to be awaiting the Sun and most other stars in the Milky Way.

However, when a star with a mass of about 1.4 to three times the mass of the Sun undergoes collapse, it will contract even further and end up as a neutron star, with a diameter of perhaps only 12.5 miles (20 km). In neutron stars, intense gravitational forces drive electrons into atomic nuclei, forcing them to combine with protons and transforming this combination into neutrons. Atomic nuclei are, therefore, obliterated in this process, and only the collective resistance of neutrons to compression halts the collapse. At this point, the star's matter is so dense that each cubic centimeter has a mass of several billion tons.

For stars that end their life having more than a few solar masses, even the resistance of neutrons is not enough to stop the unyielding gravitational collapse. In death such massive stars may ultimately become black holes—incredibly dense point masses or singularities that are surrounded by a literal black region in which gravitational attraction is so strong that nothing, not even light itself, can escape.

Currently, many scientists relate the astronomical phenomena called supernovas and pulsars with neutron stars and their evolution. The final collapse of a giant star to the neutron stage may give rise to the physical conditions that cause its outer portions to explode, creating a supernova. This type of cosmic explosion releases so much energy that its debris products will temporarily outshine all the ordinary stars in the galaxy.

A regular *nova* (the Latin word for "new," the plural of which is novas or novae) occurs more frequently and is far less violent and spectacular. One common class, called recurring novas, is due to the nuclear ignition of gas being drawn from a companion star to the surface of a white dwarf. Such binary star systems are quite common; sometimes the stars will have orbits that regularly bring them close enough for one to draw off gas from the other.

When a supernova occurs at the end of a massive star's life, the violent explosion fills vast regions of space with matter that may radiate for hundreds or even thousands of years. The debris created by a supernova explosion eventually will cool into dust and gas, become part of a giant interstellar cloud, and perhaps once again be condensed into a star or planet. Most of the heavier elements found on Earth are thought to have originated in supernovas, because the normal thermonuclear fusion processes cannot produce such heavy elements. The violent power of a supernova explosion can, however, combine lighter elements into the heaviest elements found in nature (e.g., lead, thorium, and uranium). Consequently, both the Sun and its planets were most likely enriched by infusions of material hurled into the interstellar void by ancient supernova explosions.

Pulsars, first detected by radio astronomers in 1967, are sources of very accurately spaced bursts, or pulses, of radio signals. These radio signals are so regular, in fact, that the scientists who made the first detections were startled into thinking that they might have intercepted a radio signal from an intelligent alien civilization.

The pulsar, named because its radio-wave signature regularly turns on and off, or pulses, is considered to be a rapidly spinning neutron star. One pulsar is located in the center of the Crab Nebula, where a giant cloud of gas is still glowing from a supernova explosion that occurred in the year 1054 C.E.—a spectacular celestial event observed and recorded by ancient Chinese and Korean astronomers. The discovery of this pulsar allowed scientists to understand both pulsars and supernovas.

In a supernova explosion, a massive star is literally destroyed in an instant, but the explosive debris lingers and briefly outshines everything in the galaxy. In addition to scattering material all over interstellar space, supernova explosions leave behind a dense collapsed core made of neutrons. This neutron star, with an immense magnetic field, spins many times a second, emitting beams of radio waves, X-rays, and other radiations. These radiations may be focused by the pulsar's powerful magnetic field and sweep through space much like a revolving lighthouse beacon. The neutron star, the end product of a violent supernova explosion, becomes a pulsar.

Astrophysicists must develop new theories to explain how pulsars can create intense radio waves, visible light, X-rays, and gamma rays, all at the same time. Orbiting X-ray observatories have detected X-ray pulsars that scientists believe are caused by a neutron star pulling gaseous matter from a normal companion star in a binary star system. As gas is sucked away from the normal companion to the surface of the neutron star, the gravitational attraction of the neutron star heats up the gas to millions of kelvins; this causes the gas to emit X-rays.

The advent of the space age and the use of powerful orbiting observatories, such as the *Hubble Space Telescope,* the *Compton Gamma Ray Observatory,* the *Chandra X-ray Observatory,* and the *Spitzer Space Telescope,* to view the universe as never before possible has greatly increased scientific knowledge about the many different types of stellar phenomena. Most exciting of all, perhaps, is the fact that this process of astrophysical discovery has really only just begun.

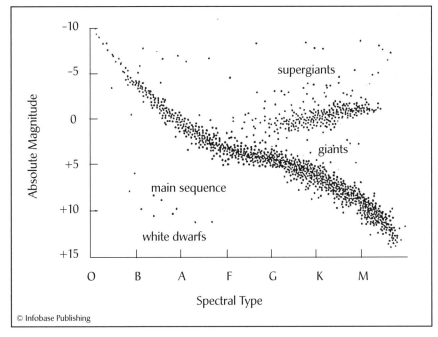

This diagram is one version of the famous Hertzsprung-Russell diagram—a useful graphic depiction of the different types of stars arranged by spectral classification and absolute magnitude (a measure of the brightness or luminosity of a star). (*NASA and the author*)

✦ Henry Norris Russell and the Hertzsprung-Russell Diagram

The American astronomer and astrophysicist Henry Norris Russell (1877–1957) was one of the most influential astronomers in the first half of the 20th century. As a student, professor, and observatory director, he worked nearly 60 years at Princeton University—a truly productive period that also included vigorous retirement activities as professor emeritus and observatory director emeritus.

Primarily a theoretical astronomer, he made significant contributions in spectroscopy and to astrophysics. Independent of Ejnar Hertzsprung (1873–1967), Russell investigated the relationship between absolute stellar magnitude and a star's spectral class. By 1913 their independent but complementary efforts resulted in the development of the famous Hertzsprung-Russell (H-R) diagram—a diagram that is of fundamental importance to all modern astronomers who wish to understand the theory of stellar evolution. Russell also performed pioneering studies of eclipsing binaries and made preliminary estimates of the relative abundance of elements in the universe. Often called "the dean of American astronomers," he served the astronomical community very well as a splendid teacher, writer, and research adviser.

Russell was born on October 25, 1877, in Oyster Bay, New York. The son of a Presbyterian minister, he received his first introduction to astronomy at age five, when his parents showed him the 1882 transit of Venus across the Sun's disk. He completed his undergraduate education at Princeton University in 1897, graduating with the highest academic honors ever awarded by that institution—namely, *insigni cum laude* (with extraordinary honor). Russell remained at Princeton for his doctoral studies in astronomy, again graduating with distinction (summa cum laude) in 1900.

Following several years of postdoctoral research at Cambridge University, he returned to Princeton in 1905 to accept an appointment as an instructor in astronomy. He remained affiliated with Princeton for the remainder of his life. He became a full professor in 1911 and the following year became director of the Princeton Observatory. He remained in these positions until his retirement in 1947. Starting in 1921, Russell also held an additional appointment as a research associate of the Mount Wilson Observatory in the San Gabriel Mountains just north of Los Angeles, California. In 1927 he received an appointment to the newly endowed C.A. Young Research Professorship at Princeton—a special honor bestowed up him by his undergraduate classmates from the Class of 1897.

Starting with his initial work at Cambridge University on the determination of stellar distances, Russell began to assemble data from different classes of stars. He noticed these data related spectral type and absolute magnitude and soon concluded (independent of Hertzsprung) that there were actually two general types of stars: giants and dwarfs. By the early 1910s, their efforts resulted in the development of the Hertzsprung-Russell diagram.

The Hertzsprung-Russell diagram is an innovative graph that characterizes a star by depicting is brightness (luminosities) as a function of temperature (spectral type). Since its introduction by Russell at a technical meeting in 1913, it has become one of the most important tools in modern astrophysics. So-called dwarf stars, including the Sun, lie on the main sequence of the H-R diagram—the well-populated region that runs from the top left to lower-right portions of this graph. Giant and supergiant stars lie in the upper-right portion of the diagram above the main-sequence band. Finally, huddled down in the lower-left portion of the H-R diagram below the main-sequence band are the extremely dense, fading cores of burned-out stars known collectively as *white dwarfs*. This term is sometimes misleading because it actually applies to a variety of compact, fading stars that have experienced gravitational collapse at the end of their life cycle.

Following his pioneering work in stellar evaluation, Russell engaged in equally significant work involving eclipsing binaries. An eclipsing binary is a binary star that has its orbital plane positioned with respect to Earth in such a way that each component star is totally or partially eclipsed by the other star every orbital period. With his graduate student, Harlow Shapley (1885–1972), Russell analyzed the light from such stars to estimate stellar masses. Later he collaborated with another assistant, Charlotte Emma Moore Sitterly (1898–1990), in using statistical methods to determine the masses of thousands of binary stars.

Before the discovery of nuclear fusion, Russell tried to explain stellar evolution in terms of gravitational contraction and continual shrinkage and used the Hertzsprung-Russell diagram as a flowchart. When Hans Albrecht Bethe (1906–2005) and other physicists began to associate nuclear-fusion processes with stellar life cycles, Russell abandoned his contraction theory of stellar evolution. However, the basic information he presented in the H-R diagram still remained very useful.

In the late 1920s, Russell performed a detailed analysis of the Sun's spectrum—showing hydrogen was a major constituent. He noted the presence of other elements, as well as their relative abundances. Extending this work to other stars, he postulated that most stars exhibit a similar, general combination of relative elemental abundances (dominated by hydrogen and helium) that became know as the Russell mixture. This work reached the same general conclusion that was previously suggested in 1925 by the British astronomer Cecilia Helena Payne-Gaposchkin (1900–79).

Russell was an accomplished teacher, and his excellent two-volume textbook, *Astronomy,* that was jointly written with Raymond Dugan and John Stewart appeared in 1926–27. It quickly became a standard text in astronomy curricula at universities around the world. His *Solar System and Its Origin* (1935) served as a pioneering guide for future research in astronomy and astrophysics. Even after his retirement from Princeton in 1947, Russell remained a dominant force in American astronomy. Honored with appointments as both an emeritus professor and an emeritus observatory directory, he pursued interesting areas in astrophysics for the remainder of his life. He died in Princeton, New Jersey, on February 18, 1957.

✦ The *Hubble Space Telescope*

The *Hubble Space Telescope* (*HST*) is a cooperative program of the European Space Agency (ESA) and NASA to operate a long-lived space-based

EDWIN POWELL HUBBLE

The American astronomer Edwin Powell Hubble revolutionized scientific understanding of the universe by proving that galaxies existed beyond the Milky Way Galaxy. In the 1920s, he made important observational discoveries that completely changed how scientists view the universe. This revolution in cosmology started in 1923, when Hubble used a Cephid variable to estimate the distance to the Andromeda Galaxy. His results immediately suggested that such spiral nebulas were actually large, distant independent stellar systems or island universes. Next he introduced a classification scheme in 1925 for such nebulas (galaxies), calling them either elliptical, spiral, or irregular. This scheme is still quite popular in astronomy. In 1929 Hubble announced that in an expanding universe, the other galaxies are receding from us with speeds proportional to their distance—a postulate now known as Hubble's law. Hubble's concept of an expanding universe filled with numerous galaxies forms the basis of modern observational cosmology.

Hubble was born in Marshfield, Missouri, on November 20, 1889. Early in his life, Hubble studied law at the University of Chicago and then as a Rhodes scholar at Oxford from 1910 to 1913. However, in 1913, he made a decision fortunate for the world of science by abandoning law and studying astronomy instead. From 1914 to 1917, Hubble held a research position at the Yerkes Observatory of the University of Chicago. There, on the shores of Lake Geneva at Williams Bay, Wisconsin, Hubble started studying interesting nebulae. By 1917 he concluded that the spiral-shaped ones (now called galaxies) were quite different than the diffuse nebulae (which are actually giant clouds of dust and gas).

Following military service in World War I, he joined the staff at the Carnegie Institute's Mount Wilson Observatory, located in the San Gabriel Mountains northwest of Los Angeles. When Hubble arrived in 1919, this observatory's 100-inch- (2.54-m-) diameter telescope was the world's biggest optical instrument. Except for other scientific work during World War II, Hubble remained affiliated with this observatory for the remainder of his life. Once at Mount Wilson, Hubble used its large instrument to resume his careful investigation of nebulae.

In 1923 Hubble discovered a Cepheid variable star in the Andromeda nebula—a celestial object now known to astronomers as the Andromeda Galaxy, or M31. A Cepheid variable is one of a group of important, very bright supergiant stars that pulsate periodically in brightness. By carefully studying this particular Cepheid variable in M31, Hubble was able to conclude that it was very far away and belonged to a separate collection of stars far beyond the Milky Way. This important discovery provided the first tangible observational evidence that galaxies existed beyond the Milky Way. Through Hubble's pioneering efforts, the size of the known universe expanded by incredible proportions.

Hubble continued to study other galaxies and in 1925 introduced the well-known classification system of spiral galaxies, barred spiral galaxies, elliptical galaxies, and irregular galaxies. Then, in 1929, Hubble investigated the recession velocities of galaxies (i.e., the rate at which galaxies are moving apart) and their distances away. He discovered that the more distant galaxies were receding (going away) faster than the galaxies closer to us. This very important discovery revealed that the universe was expanding. Hubble's work provided the first observational evidence that supported big bang cosmology. Today the concept of the uni-

In the direction of the constellation Canis Major, two spiral galaxies pass by each other like majestic ships in the night. The near-collision was caught in this image taken by the Wide Field/Planetary Camera 2 of the *Hubble Space Telescope* in 2004. The larger and more massive galaxy (on the left in the image) is cataloged as NGC 2207, and the smaller galaxy (on the right) is called IC 2163. Strong tidal forces from NGC 2207 have distorted the shape of IC 2163, flinging out stars and gas into long streamers that stretch out a 100,000 light-years toward the right-hand side of the image. *(NASA, ESA, and the Hubble Heritage Team [STScI])*

verse expanding at a uniform, steady rate is codified in a simple mathematical relationship called Hubble's law in his honor. (Observations made in the late 1990s indicate that the rate of expansion is actually increasing and not uniform, throwing the world of astronomy and cosmology into a contemporary quandary.)

Hubble's law describes the expansion of the universe as a linear, steady rate. As Hubble initially observed and subsequent astronomical studies (up to the late 1990s) confirmed, the apparent reces-sion velocity (v) of galaxies is proportional to their distance (r) from an observer. The proportionality constant is H_0, the Hubble constant. Currently, proposed values for H_0 fall between 50 and 90 kilometers per second per megaparsec (km s^{-1} Mpc^{-1}). The inverse of the Hubble constant ($1/H_0$) is called the Hubble time. Astronomers use the value of the Hubble time as one measure of the age of the universe. If H_0 has a value of 50 km s^{-1} Mpc^{-1}, then the universe is about 20 billion years old (which appears far too old by recent astrophysical

(continues)

(continued)

estimates). An H_0 value of 80 km s^{-1} Mpc^{-1} suggests a much-younger universe, ranging in age between 8 and 12 billion years old.

Despite space-age investigations of receding galaxies, there is still much debate today within the astrophysical community as to the proper value of the Hubble constant. Cosmologists and astrophysicists using several techniques currently suggest that the universe is about 13.7 billion years old. Astrophysical observations of distant supernovas made in the late 1990s now suggest that rate of expansion of the universe is actually increasing. This will influence the linear expansion-rate hypothesis inherent in Hubble's law and will also influence the age of the universe estimates based on the Hubble time. However, these recent observations, while very significant and important, take little away from Hubble's brilliant pioneering work, which introduced the modern cosmological model of an enormously large, expanding universe filled with billions of galaxies.

Hubble, much like the astronomical pioneers before him, created a revolution in our understanding of the physical universe. Through his dedicated observational efforts, scientists learned about the existence of galaxies beyond the Milky Way and the fact that the universe appears to be expanding. Quite fittingly, NASA named the *Hubble Space Telescope* after him. This powerful orbiting astronomical observatory continues the fine tradition of extragalactic investigation started by Hubble, who died on September 28, 1953, in San Marino, California.

optical observatory for the benefit of the international astronomical community. The orbiting facility is named for American astronomer Edwin Powell Hubble, who revolutionized scientific knowledge of the size, structure and makeup of the universe through his pioneering observations in the first half of the 20th century.

Astronomers and space scientists are using the *HST* to observe the visible universe to distances never before obtained and to investigate a wide variety of interesting astronomical phenomena. In 1996, for example, *HST* observations revealed the existence of approximately 50 billion (10^9) more galaxies than scientists previously thought existed.

The *HST* is 43 feet (13.1 m) long and has a diameter of 14 feet (4.27 m). This space-based observatory is designed to provide detailed observational coverage on the visible, near-infrared, and ultraviolet portions of the electromagnetic spectrum. The *HST* power-supply system consists of two large solar panels (unfurled on orbit), batteries and power-conditioning equipment. The 24,000-pound- (11,000-kg-) mass free-flying astronomical observatory was initially placed into a 375-mile (600-km) low Earth orbit during the STS-31 space shuttle *Discovery* mission on April 25, 1990.

The years since the launch of *HST* have been momentous, including discovery of a spherical aberration in the telescope optical system—a flaw that severely threatened the space-based telescope's usefulness. A practical solution was found, however, and the STS-61 space shuttle *Endeavour* mission (December 1993) successfully accomplished the first on-orbit servicing and repair of the telescope. The effects of the spherical aberration were overcome, and the *HST* was restored to full

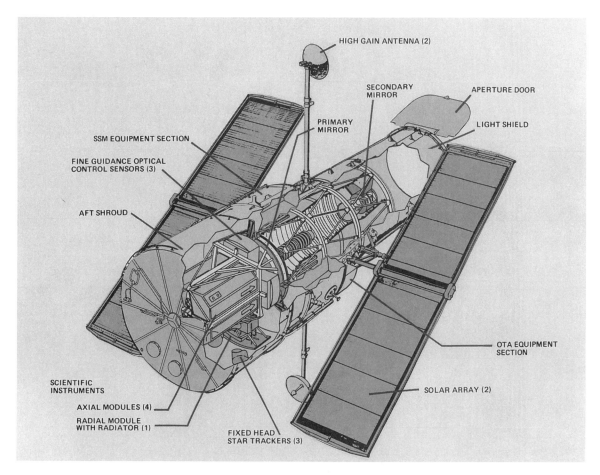

This cutaway diagram shows the overall structural configuration of NASA's *Hubble Space Telescope* (*HST*) (ca. 1985). Astronauts on board the space shuttle *Discovery* deployed this powerful optical telescope into orbit around Earth in April 1990, as part of the STS-31 mission. Named in honor of the American astronomer Edwin Powell Hubble (1889–1953) the *HST* has made significant contributions to the study of the universe. Since its initial deployment, on-orbit servicing missions by the space shuttle have permitted the exchange of scientific instruments, as well as the replacement of aging or failed spacecraft equipment, such as solar arrays and gyroscopes. (*NASA/MSFC*)

The photograph shows the *Hubble Space Telescope* being lifted into the vertical position as part of launch preparations at the Kennedy Space Center, Florida (October 1989). In April 1990, the large, free-flying optical telescope was successfully carried into low Earth orbit and deployed by the crew of space shuttle *Discovery*, as part of the STS–31 mission. *(NASA/KSC)*

functionality. During this mission, the astronauts changed out the original Wide Field/Planetary Camera (WF/PC1) and replaced it with the Wide Field/Planetary Camera 2 (WF/PC2). The relay mirrors in WF/PC2 are spherically aberrated to correct for the spherically aberrated primary mirror of the observatory. *Hubble's* primary mirror is two micrometers too flat at the edge, so corrective optics within WF/PC2 are too high by that same amount. In addition, a corrective optics package, called the Corrective Optics Space Telescope Axial Replacement, or COSTAR, replaced the High Speed Photometer during the STS-61 servicing mission. COSTAR is designed to optically correct the effect of the primary mirror's aberration on the telescope's three other scientific instruments: the Faint Object Camera (FOC), built by the ESA; Faint Object Spectrograph (FOS); and the Goddard High Resolution Spectrograph (GHRS).

In February 1997, a second servicing operation (*HST* SM-02) was successfully accomplished by the STS-82 space shuttle *Discovery* mission. As part of this *HST* servicing mission, astronauts installed the Space Telescope Imaging Spectrograph (STIS) and the Near-Infrared Camera and Multi-Object Spectrometer (NICMOS). These instruments replaced the GHRS an the FOS, respectively. The STIS observes the universe in four spectral bands that extend from the ultraviolet through the visible and into the near infrared. Astronomers use the STIS to analyze the temperature, composition, motion, and other important properties of celestial objects. Operating at near-infrared wavelengths, the NICMOS is letting astronomers observe the dusty cores of active galactic nuclei galaxies and investigate interesting protoplanetary disks around stars.

On November 13, 1999, flight controllers placed the *Hubble Space Telescope* in a safe mode after the failure of a fourth gyroscope. In this safe mode, the *HST* could not observe celestial targets, but the observatory's overall safety was preserved. This protective mode allowed ground control of the telescope, but with only two gyroscopes working, *HST* could not point with the precision necessary for scientific observations. Flight controllers on the ground also closed the telescope's aperture door in order to protect the optics and then aligned the spacecraft to make sure that *Hubble Space Telescope's* solar panels would receive adequate illumination from the Sun.

On December 19, 1999, seven astronauts boarded the space shuttle *Discovery* and departed the Kennedy Space Center to make an on-orbit house call to the stricken observatory. When the third of the *HST's* gyroscopes failed (the space-based telescope needs at least three to point with scientific accuracy), NASA made a decision to split the third servicing mission into two parts: Servicing Mission 3A (SM3A)

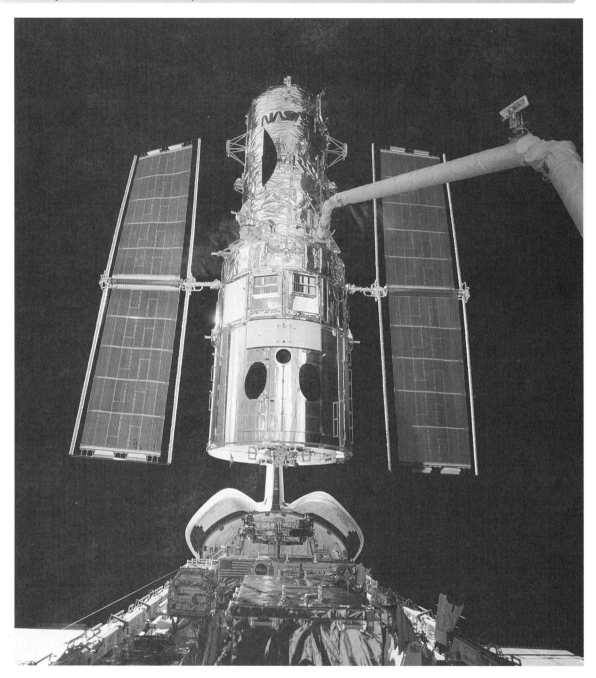

NASA's *Hubble Space Telescope* (*HST*) is shown here being carefully lifted up out of the payload bay of the space shuttle *Discovery* and then placed into sunlight by the shuttle's robot arm in February 1997. This event took place during the STS–82 mission, which NASA also refers to as the second *HST* servicing mission (*HST* SM-02). (*NASA/JSC*)

and Servicing Mission 3B (SM3B). Six days and three extravehicular activities later, the STS-103 mission crew had replaced worn or out-dated equipment and performed critical maintenance during SM3A. The most important task was to replace the gyroscopes that accurately point the telescope at celestial targets. The astronauts also installed an advanced central computer, a digital data recorder, battery improvement kits, and new outer layers of thermal protection material. After the *Discovery* and its crew returned safely to Earth (on December 27), the *HST* flight controllers turned on the orbiting instrument. Because of this successful repair mission, the *Hubble Space Telescope* was as good as new and returned to its astronomical duties.

The second part of the third servicing mission took place in March 2002. NASA launched the space shuttle *Columbia* so its crew of seven astronauts could rendezvous with the *HST* and perform a series of upgrades. The STS-109 mission is also known as SM3B. During the mission, which started on March 1, 2002, the astronauts performed five extravehicular activities, or space walks. Their primary task was to install a new science instrument called the Advanced Camera for Surveys (ACS). The ACS brought the nearly 12-year-old orbiting telescope into the 21st century. With its wide field of view, sharp image quality, and enhanced sensitivity, the ACS doubled the telescope's field of view and collected data 10 times faster than the Wide Field/Planetary Camera 2—the *HST*'s earlier surveying instrument.

During this servicing mission, the astronauts also upgraded the telescope's electric power system. They replaced the spacecraft's four eight-year-old solar panels with smaller, rigid ones that produce 30 percent more electricity. During the final extravehicular activity, the STS-109 astronauts installed a new cooling system for the Near-Infrared Camera and Multi-Object Spectrometer, which became inactive when its 220-pound (100-kg) block of nitrogen ice coolant was depleted. The new refrigeration system chills NICMOS's infrared detectors to below −315 °F (−193 °C). Finally, the shuttle astronauts replaced one of the four reaction-wheel assemblies that make up the *HST*'s pointing control system.

On January 16, 2004, NASA Administrator Sean O'Keefe announced his decision to cancel the fourth servicing mission—the last scheduled flight of the space shuttle to the *Hubble Space Telescope*. Astronauts on this servicing mission (planned for 2006) would have performed maintenance work and installed new instruments. New shuttle flight–safety guidelines, following the loss of the *Columbia* (on February 1, 2003), weighed heavily in the administrator's decision. However, NASA's current administrator, Michael Griffin, has decided to pursue a space shuttle servicing mission to the *Hubble Space Telescope* as part of the space agency's fiscal year 2007 budget plan. In addition a robotic mission to capture the *HST* and then safely de-orbit the large structure at the end of its useful scientific life is also under consideration (as a backup). NASA flight managers cannot project *Hubble's* lifetime beyond 2007, because that is when critical components on the telescope will require replacement or repair. Two items of special concern are the telescope's gyroscopes and batteries. The replacement for the *Hubble Space Telescope* is a space-based observatory called the *James Webb Space Telescope*—now scheduled for launch in August 2011.

Although *HST* operates around the clock, not all of its time is spent observing the universe. Each orbit lasts about 95 minutes, with time allocated for housekeeping functions and for observations. Housekeeping functions include turning the telescope to acquire a new target, or to avoid the Sun or the Moon; switching communications antennae and data transmission modes; receiving command loads and downloading data; calibrating; and similar activities. Responsibility for conducting and coordinating the science operations of the *Hubble Space Telescope* rests with the Space Telescope Science Institute on the Johns Hopkins University Homewood Campus in Baltimore, Maryland.

Because of *HST's* location above the Earth's atmosphere, its science instruments can produce high-resolution images of astronomical objects. Ground-based telescopes, influenced by atmospheric effects, can seldom provide resolution better than 1.0 arc-second, except perhaps momentarily under the very best observing conditions. The *Hubble Space Telescope's* resolution is about 10 times better, or 0.1 arc second.

Here are just a few of the exciting discoveries provided by the *Hubble Space Telescope:* (1) It gave astronomers their first detailed view of the shapes of 300 ancient galaxies located in a cluster 5 billion light-years away; (2) it provided astronomers their best look yet at the workings of a black-hole "engine" in the core of the giant elliptical galaxy NGC4261, located 45 million light-years away in the constellation Virgo; (3) *HST's* detailed images of newly forming stars helped confirm more than a century of scientific hypothesis and conjecture on how a solar system begins; and (4) *HST* gave astronomers their earliest look at a rapidly ballooning bubble of gas blasted off a star (Nova Cygni, which erupted February 19, 1992).

Pushing the limits of its powerful vision, the *HST* has also uncovered the oldest burned-out stars in the Milky Way Galaxy. Detection of these extremely old, dim, white dwarfs have provided astronomers a completely independent indication of the age of the universe—an assessment that does not rely on measurements of the universe's expansion. The ancient white dwarfs, as imaged by the *Hubble Space Telescope* in 2001, appear to be between 12 to 13 billion years old. Finding these oldest stars puts astronomers well within arm's reach of calculating the absolute age of the universe, since the faintest and coolest white dwarfs within globular clusters can yield a globular cluster's age, and globular clusters are among the oldest clusters of stars in the universe.

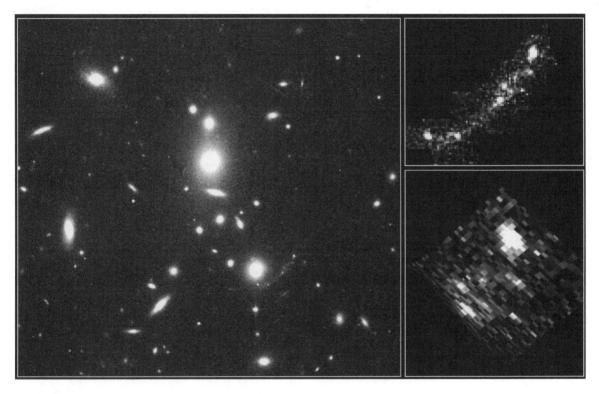

The *Hubble Space Telescope* image (on the left) of the galaxy cluster CL1358+62 has uncovered a gravitationally lensed image of a more distant galaxy that is located far beyond the cluster. The image on the upper right is a close-up look at the gravitationally lensed image of the distant galaxy. The stretched-out image reveals tiny knots of vigorous star birth activity. By collecting this image (in 1997), the *Hubble Space Telescope* has provided astronomers their first detailed look at the early construction phase of a galaxy undergoing formation. The image in the lower right is a theoretical model of the cluster that astronomers used to "unsmear" the gravitationally lensed image back into the galaxy's normal appearance. Astronomers estimate the gravitationally lensed young galaxy is about 13 billion years away, while the foreground cluster responsible for the gravitational lensing is about 5 billion light-years from Earth. *(NASA, Marijn Franx [University of Groningen, The Netherlands], and Garth Illingworth [University of California, Santa Cruz])*

Finally, from September 2003 to January 2004, the *Hubble Space Telescope* performed the Hubble Ultra Deep Field (HUDF) observation—the deepest portrait of the visible universe ever achieved by humankind. The HUDF revealed the first galaxies to emerge from the so-called dark ages—the time shortly after the big bang event when

This *Hubble Space Telescope* image (taken in 2004) shows thousands of galaxies and is the deepest visible-light image of the cosmos yet taken. Called the Hubble Ultra Deep Field (HUDF), the galaxy-studded view represents a deep core sample of the universe, cutting across billions of light-years. The cosmic snapshot includes galaxies of various ages, sizes, shapes, and colors. The smallest and reddest galaxies appearing in the image may be among the most distant known, existing when the universe was just 800 million years old. The nearest galaxies—the larger, brighter, well-defined spirals and ellipticals—thrived about 1 billion years ago (when the universe was about 13 billion years old). *(NASA, ESA, S. Beckwith [STScI] and the HUDF Team)*

the first stars reheated the cold universe. This historic portrait of the early universe involved two separate long-exposure images captured by the *HST* ACS and NICMOS.

Imagery from the *HST* has revolutionized the scientific view of the visible universe. Refurbished by shuttle-servicing missions, this amazing facility should continue to provide astrophysicists and astronomers with incredibly interesting data for several more years.

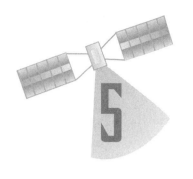

Gamma-ray Astronomy and the *Compton Gamma Ray Observatory*

Gamma-ray astronomy is the youngest branch of high-energy astronomy and astrophysics. The discipline is based on the detection of the energetic gamma rays associated with supernovas, exploding galaxies, quasars, pulsars, and phenomena near suspected black holes.

Gamma rays (symbol: γ) are extremely high-energy, very-short-wavelength packets or quanta of electromagnetic radiation. Gamma-ray photons are similar to X-rays, except that they are usually more energetic and originate from processes and transitions within the atomic nucleus. Gamma rays typically have energies between 10,000 electron volts and 10 million electron volts (i.e., between 10 keV and 10 MeV), with correspondingly short wavelengths and high frequencies. The processes associated with gamma-ray emissions in astrophysical phenomena include: (1) the decay of radioactive nuclei, (2) cosmic-ray interactions, (3) curvature radiation in extremely strong magnetic fields, and (4) matter-antimatter annihilation. Gamma rays are very penetrating and are best stopped or shielded against by dense materials, such as lead or tungsten.

A hypernova is a new type of supernova that some astrophysicists suggest may be connected to the mysterious and very energetic phenomenon of gamma-ray bursts. In 2003 astronomers fortuitously observed an energetic gamma-ray burst, called GRB 030329, in the X-ray, optical, and radio wavelengths. These recent data provided scientists with an important clue in their efforts to unravel one of the most mysterious phenomena in the universe.

✧ Gamma-ray Bursts

A gamma-ray burst (GRB) is an outburst that radiates tremendous amounts of energy equal to or greater than a supernova, in the form of gamma rays and X-rays. These bursts take place in less than a few minutes.

The gamma-ray burster, as the intriguing phenomenon is sometimes called, is one of the greatest astronomical mysteries.

Starting in the later 1960s, astronomers observing the heavens at very short wavelengths began detecting these incredibly brief and intense bursts of gamma rays from seemingly random locations in the sky. They observed that a few times a day, the sky would light up with an incredible flash, or burst, of gamma rays. Often the burst outshone all of the other sources of cosmic radiation added together. The source of the burst then disappeared completely. Even more puzzling, no one was able to predict when the next burst would occur or from what direction in the sky it would come.

Gamma-ray bursts were first discovered in the late 1960s by a scientist named Ray Klebesadel who was working at the Los Alamos National Laboratory (LANL). His job involved analyzing data from a series of Vela nuclear test–detection satellites, then being operated by the U.S. Air Force. Klebesadel and his colleagues were examining stacks of computer printout in an effort to make sure the gamma-ray detectors and other nuclear treaty–monitoring instruments constructed by the Department of Energy for use by the Department of Defense on this series of military satellites, were working properly.

The U.S. Air Force launched Vela nuclear-detonation-detection satellites in pairs into 68,365-mile (110,000-km) circular orbits around Earth. Shown here are the *Vela 5A* and *Vela 5B* spacecraft prior to launch in May 1969 by a Titan IIIC rocket. These spin-stabilized, polyhedral satellites were specifically designed and operated solely for detecting nuclear explosions on or above Earth's surface and in space. *(U.S. Air Force)*

In mid-1969 Klebesadel was examining archived Vela gamma-ray data recorded on July 2, 1967. He noticed a spike in the data, a dip, a second spike, and a long tail off. This was not the telltale gamma-ray signature of a clandestine nuclear test in outer space. Klebesadel and his colleagues at Los Alamos and the U.S. Air Force scientists with whom they were working all wondered what the very unusual data could represent.

After this first unusual gamma-ray event was noticed, other similar events were quickly recognized in the archived Vela spacecraft data. With the timing between *Vela 5* and *Vela 6* synchronized to within ¹⁄₆₄ of a second, the Vela team was able to triangulate the locations of the bursts by comparing differences in the arrival times at widely separated satellites. The American scientists confirmed their suspicion that the gamma-ray bursts came from outside the solar system. The random scatter of these events across the sky also suggested to them that the sources were most likely outside of the Milky Way Galaxy. By 1973, when Klebesadel and his team were ready to publish the results in *Nature* and present them at the American Astronomical Society meeting, there were at least 16 confirmed bursts. National security restrictions at the time delayed release of the data and prevented any direct reference to U.S. Air Force involvement. With the end of the cold war in 1989, those restrictions have since been removed.

Following the discovery of gamma-ray bursts in the late 1960s and the public announcement in 1973, astronomers used the best-available space-based gamma-ray detection instruments to construct a catalog of these mysterious events. As the number of observed GRBs grew, many theories emerged within the international astronomical community concerning their origin. Scientists also argued among themselves as to whether the GRBs were taking place in the Milky Way Galaxy or in other galaxies. Unfortunately, the addition of each newly observed GRB tended to reveal little more than the fact that GRBs never repeated from the same cosmic source or location.

Then NASA's *Compton Gamma Ray Observatory*—launched in 1991—provided a wealth of new GRB observations. The spacecraft's burst and transient source experiment (BATSE) was capable of monitoring the sky with unprecedented sensitivity. One thing soon became clear as the catalog of GRBs observed by BATSE grew—the GRBs were in no way correlated with sources in the Milky Way Galaxy. In 1997 the Italian-Dutch *BeppoSAX* satellite made a breakthrough in scientists' basic understanding of gamma-ray bursts. Using a particularly effective combination of gamma-ray and X-ray telescopes, *BeppoSAX* detected afterglows from a few GRBs and precisely located the sources in a way that allowed other telescopes (sensitive to different portions of the electromagnetic spectrum) to promptly study the same location in the sky. This effort showed astronomers that GRBs are produced in very distant galaxies—requiring that the explosions producing the gamma-ray bursts be incredibly powerful.

The next important breakthrough in GRB understanding occurred on January 23, 1999, when an enormously powerful event (called GRB990123) was detected. Alerted by sophisticated space-based detectors, astronomers were able to quickly observe this event with an unprecedented range of wavelengths and timing sensitivities. The event was very far away. If astronomers assumed isotropic emission of the gamma rays, this powerful GRB would have involved the release of the energy equivalent of twice the rest mass energy of a neutron star. If, on the other hand, astronomers assumed that the emitted energy was really being beamed out of this distant GRB in a preferred direction (here one that just happened to point directly toward Earth), then the required energies became more reasonable and a little easier to explain.

Today astronomers *tenuously* suggest that gamma-ray bursts are produced by materials shooting toward Earth at nearly the speed of light. The material in question is ejected during the collision of two neutron stars or black holes. An alternate speculation is that the GRBs arise from a hypernova—the huge explosion hypothesized to occur when a supermassive star ends its life and collapses into a black hole. However, astronomers also recognize that many more multi-wavelength, prompt observations of future GRBs are required before they can confidently model the central engine (or engines, since there may be more than one mechanism) that produces these powerful and mysterious gamma-ray bursts.

✧ NASA's *Compton Gamma Ray Observatory*

The *Compton Gamma Ray Observatory* (*CGRO*) was one of the four Great Observatories developed by NASA to support space-based astronomy. The other three observatories in this special, scientific spacecraft family are the *Chandra X-ray Observatory,* the *Hubble Space Telescope* and the *Spitzer Space Telescope.*

The *CGRO* was deployed successfully into low-Earth-orbit (LEO) by the crew of space shuttle *Atlantis* on April 7, 1991, during shuttle mission STS-37. The observatory was then boosted to a higher-altitude circular orbit, where it could accomplish its scientific mission. The large, 35,860-pound- (16,300-kg-) mass spacecraft carried a variety of sensitive instruments designed to detect gamma rays over an extensive range of energies from about 30 kiloelectron volts (keV) to 30 billion electron volts (GeV). The *CGRO* was an extremely powerful tool for investigating some of the most puzzling astrophysical mysteries in the universe, including energetic gamma-ray bursts, pulsars, quasars, and active galaxies. NASA named this spacecraft in honor of the American physicist Arthur Holly Compton. At the end of

NASA's massive *Compton Gamma Ray Observatory* being deployed into low Earth orbit by the crew of the space shuttle *Atlantis,* during the STS-35 mission on April 1991. *(NASA)*

ARTHUR HOLLY COMPTON

The American physicist Arthur Holly Compton was one of the pioneers of high-energy physics. In 1927 he shared the Nobel Prize in physics for his investigation of the scattering of high-energy photons by electrons—an important phenomenon now called the Compton effect. His research efforts in 1923 provided the first experimental evidence that electromagnetic radiation possessed both particle-like and wavelike properties. Compton's important discovery made quantum physics credible.

Compton was born on September 10, 1892, into a distinguished intellectual family in Wooster, Ohio. His father was a professor at Wooster College, and his older brother (Karl) studied physics and went on to become the president of the Massachusetts Institute of Technology. As a youth, Arthur Compton experienced two very strong influences from his family environment: a deep sense of religious service and the noble nature of intellectual work.

Compton carefully weighed his career options and then followed his family's advice by selecting a career of service in physics. Upon completion of his undergraduate degree at Wooster College in 1913, he joined his older brother at Princeton. He subsequently received his master of arts degree in 1914 and his Ph.D. degree in 1916 from Princeton University. For his doctoral research, Compton studied the angular distribution of X-rays reflected from crystals.

After spending a year as a physics instructor at the University of Minnesota, Compton worked for two years in Pittsburgh, Pennsylvania, as an engineering physicist with the Westinghouse Lamp Company. Then, in 1919, he received one of the first National Research Council fellowships awarded by the American government. Compton used this fellowship to study gamma-ray scattering phenomena at Baron Ernest Rutherford's Cavendish Laboratory in England. While working with Rutherford, he verified the puzzling results obtained by other physicists—namely, that gamma rays experienced a variation in wavelength as a function of scattering angle.

The following year, Compton returned to the United States to accept a position as head of the department of physics at Washington University in Saint Louis, Missouri. There, working with X-rays, he resumed his investigation of the puzzling mystery of photon scattering and wavelength change. By 1922 his experiments revealed that there definitely was a measurable shift of X-ray (photon) wavelength with scattering angle—a phenomenon now called the Compton effect. He applied special relativity and quantum mechanics to explain the results, presented in his famous paper "A Quantum Theory of the Scattering of X-rays by Light Elements," which appeared in the May 1923 issue of the *Physical Review*. In 1927 Compton shared the Nobel Prize in physics with Charles Wilson (1869–1959) for his pioneering work on the scattering of high-energy photons by electrons. (Wilson received his share of that year's prestigious award in physics for his invention of the cloud chamber.)

It was Wilson's cloud chamber that helped Compton verify the behavior of X-ray scattered recoiling electrons. Telltale cloud tracks of recoiling electrons provided Compton his corroborating evidence of the particle-like behavior of electromagnetic radiation. His precise experiments depicted the increase in wavelength of X-rays due to the scattering of the incident radiation by free electrons. Since Compton's results implied that the scattered X-ray photons had less energy than the original X-ray photons, he became the first scientist to experimentally demonstrate the particle-like quantum nature of electromagnetic waves. His book *Secondary Radiations Produced by X-rays*, published in 1922, described much of this important research and his experimental procedures. The discovery of Compton scattering (as the Compton effect is also called) served as the technical catalyst for the acceptance and rapid development of quantum mechanics in the 1920s and 1930s.

Compton scattering is the physical principle behind many of the advanced X-ray and gamma-ray detection techniques used in contemporary high-energy astrophysics. In recognition of his uniquely important contributions to modern astronomy, the National Aeronautics and Space Administration named a large orbiting high-energy astrophysics observatory the *Compton Gamma Ray Observatory* in his honor.

In 1923 Compton became a physics professor at the University of Chicago. Once settled in at the new campus, he resumed his world-changing research with X-rays. An excellent teacher and experimenter, he wrote the 1926 textbook *X-rays and Electrons* to summarize and propagate his pioneering research experiences. From 1930 to 1940, Compton led a worldwide scientific study to measure the intensity of cosmic rays and to determine any geographic variation in their intensity. His precise measurements showed that cosmic-ray intensity actually correlated with geo-magnetic latitude rather than geographic latitude. Compton's results implied that cosmic rays were very energetic, charged particles interacting with Earth's magnetic field. His pre–space age efforts became a major contribution to space physics and stimulated a great deal of scientific interest in understanding the Earth's magnetosphere—an interest that gave rise to many of the early satellite payloads, including James Van Allen's instruments on the American *Explorer 1* satellite.

During World War II, Compton played a major role in the development and use of the American atomic bomb. He served as a senior scientific adviser and was also the director of the Manhattan Project's Metallurgical Laboratory (Met Lab) at the University of Chicago. Under Compton's leadership in the Met Lab program, the brilliant Italian-American physicist Enrico Fermi (1901–54) was able to construct and operate the world's first nuclear reactor on December 2, 1942. This successful uranium-graphite reactor, called Chicago Pile One, became the technical ancestor for the large plutonium-production reactors built at Hanford, Washington. The Hanford reactors produced the plutonium used in the world's first atomic explosion, the Trinity device detonated in southern New Mexico on July 16, 1945, and also in the Fat Man atomic weapon dropped on Nagasaki, Japan, on August 9, 1945.

Following World War II, Compton put aside physics research and accepted the position of chancellor at Washington University in Saint Louis, Missouri. He served the university well as its chancellor until 1953 and then continued his relationship as a professor of natural philosophy, until failing health forced him to retire in 1961. He died in Berkeley, California, on March 15, 1962.

its useful scientific mission, flight controllers intentionally commanded the massive spacecraft to perform a de-orbit burn. This action caused the *CGRO* to reenter Earth's atmosphere in June 2000 and safely plunge into a remote region of the Pacific Ocean.

The *CGRO* carried a complement of four instruments that provided simultaneous observations, covering five decades of gamma-ray energy from 30,000 electron volts (30 keV) to 30 billion electron volts (30 GeV). In order of increasing spectral energy coverage, they were the burst and transient source experiment (BATSE), the oriented scintillation spectrometer experiment (OSSE), the imaging Compton telescope (COMPTEL), and the energetic gamma-ray experiment telescope (EGRET). For each of the instruments, an improvement in sensitivity of better than a factor of ten was realized over previous missions.

The four *CGRO* instruments were much larger and more sensitive than any gamma-ray telescopes previously flown in space. The large size was necessary because the number of gamma-ray interactions that can be recorded is directly related to the mass of the detector. Since the number of gamma-ray photons from celestial sources is very small compared to the number of optical photons, astrophysicists must use large instruments to detect a significant number of gamma rays in a reasonable amount of time. The combination of these instruments could detect photon energies from about 30 keV to more than 30 GeV.

An appreciation of the purpose and design of the *CGRO*'s four instruments can be gained from understanding that above the typical energies of X-ray photons (~10 keV, or about 10,000 times the energy of optical photons), materials cannot easily refract or reflect the incoming radiation to form a picture. As a consequence, scientists had to use alternative methods to collect gamma-ray photons and thereby form images of gamma-ray sources in the sky. At gamma-ray energies, astrophysicists elected to use three methods, sometimes in combination: (1) partial or total absorption of the gamma ray's energy within a high-density medium, such as a large crystal of sodium iodide; (2) collimation using heavy absorbing material, to block out most of the sky and create a small field of view; and (3) at sufficiently high energies, use of the conversion process from gamma rays to electron-positron pairs in a spark chamber, which leaves a telltale directional signature of the incoming photon.

The *Compton Gamma Ray Observatory* had a diverse scientific agenda, which included studies of very-energetic celestial phenomena: solar flares, gamma-ray bursts, pulsars, nova and supernova explosions, accreting black holes of stellar mass, quasar emission, and interactions of cosmic rays with the interstellar medium. Scientists have made many exciting discoveries using the *CGRO*'s instruments, some previously anticipated and some completely unexpected. For example, they discovered that the all-sky map produced by EGRET was dominated by emission from interactions between cosmic rays and the interstellar gas along the plane of the Milky Way Galaxy. Some point sources in this map are pulsars along the plane. At least seven pulsars are now known to emit in the gamma-ray portion of the spectrum, and five of these gamma-ray pulsars have been discovered since the *CGRO* was launched. One of the major discoveries made by EGRET was the class of objects known as blazars—quasars that emit the majority of their electromagnetic energy in the 30 MeV to 30 GeV portion of the electromagnetic spectrum. Blazars, which are at cosmological distances, have sometimes been observed to vary on timescales of days.

An all-sky map made by COMPTEL demonstrated the power of imaging in a narrow band of gamma-ray energy. This particular map revealed unexpectedly high concentrations of radioactive aluminum-26 in small regions. A COMPTEL (gamma-ray) image made several interesting high-energy objects "visible," including two pulsars, a flaring black-hole candidate and a gamma-ray blazar. In another map of the galactic center region made by OSSE, the instrument's scanning observations revealed gamma-ray radiation from the annihilation of positrons and electrons in the interstellar medium. The spectrum of a solar flare recorded by OSSE gave scientists direct evidence of accelerated particles smashing into material on the Sun's surface, exciting nuclei that then radiated in gamma rays.

One of BATSE's primary objectives was the study of the mysterious phenomenon of gamma-ray bursts—brief flashes of gamma rays that occur at unpredictable locations in the sky. BATSE's all-sky map of burst positions showed that, unlike galactic objects, which cluster near the plane or center of the Milky Way Galaxy, these bursts come from all directions. Through the use of *CGRO* data, astrophysicists have now established a cosmological origin for gamma-ray bursts—that is, one well beyond the Milky Way Galaxy. Burst light curves suggest that a chaotic phenomenon is at work; no two have ever appeared exactly the same. An average light curve for bright and dim gamma-ray bursts appears consistent with the current explanation that these bursts take place at cosmological distances: The dim ones, which presumably are farther away, are stretched more in cosmic time than are the bright ones, as the events participate in the general expansion of the universe.

The *CGRO* spacecraft was a three-axis, stabilized, free-flying Earth satellite capable of pointing at any celestial target for a period of 14 days or more with an accuracy of 0.5 degrees. Absolute timing is accurate to 0.1 millisecond. This important orbiting laboratory had an onboard propulsion system with approximately 4,090 pounds (1,860 kg) of monopropellant hydrazine for orbit maintenance. At the end of *CGRO*'s scientific mission (in June 2000), flight controllers used an intentionally reserved portion of this propellant supply to have the spacecraft successfully execute a carefully planned de-orbit burn with a subsequent controlled reentry into a preselected, remote area of the Pacific Ocean.

✧ *Cosmic Ray Satellite (COS-B)*

The European Space Agency (ESA) developed the *Cosmic Ray Satellites* (*COS-B*) to study extraterrestrial gamma radiation in the energy range from about 25 MeV to 5 GeV region. *COS-B* was launched by a Delta expendable rocket vehicle on August 9, 1975, from Vandenberg Air Force Base in California and successfully placed in a highly elliptical around Earth.

The COS-B satellite's highly eccentric orbit had a periapsis of 211 miles (340 km), an apoapsis of 62,075 miles (99,876 km), a period of 37.1 hours, an inclination of 90.13 degrees, and an eccentricity of 0.881. The gamma-ray telescope onboard this satellite provided scientists with their first detailed view of the Milky Way Galaxy in the gamma-ray portion of the electromagnetic spectrum. Originally projected to operate for two years, the spacecraft operated successfully for six years and eight months (from August 1975 to April 1982).

✧ *Swift* Spacecraft

NASA's *Swift* spacecraft lifted off aboard a Boeing Delta II rocket from pad 17-A at Cape Canaveral Air Force Station, Florida, on November 20, 2004. Following this successful launch, *Swift* started its mission to study gamma-ray bursts and to identify their origins. *Swift* is a first-of-its-kind, multi-wavelength space-based observatory dedicated to the study of gamma-ray bursts (GRBs).

The scientific spacecraft has three instruments that work together to observe GRBs and afterglows in the gamma-ray, X-ray, ultraviolet, and optical wave bands. The main mission objectives for *Swift* are to determine the origin of gamma-ray bursts, classify gamma-ray bursts and search for new types, determine how the GRB blast wave evolves and interacts with the surroundings, use gamma-ray bursts to study the early universe, and perform the first sensitive hard X-ray survey of the sky. *Swift*'s three telescopes span the gamma-ray, X-ray, ultraviolet, and optical light bands. This represents a swath of the electromagnetic spectrum over a million

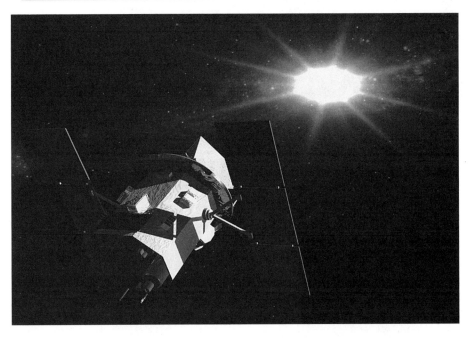

An artist's rendering depicting NASA's *Swift* spacecraft collecting data from the mysterious astronomical phenomenon known as a gamma-ray burst. *(NASA)*

times wider than what the *Hubble Space Telescope* detects. NASA scientists anticipate that, during its planned two-year mission, *Swift* will observe more than 200 gamma-ray bursts—the most comprehensive study of GRB afterglows to date.

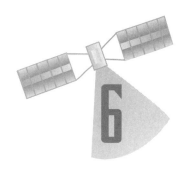

X-ray Astronomy and the *Chandra X-ray Observatory*

X-ray astronomy is the most advanced of the three general disciplines associated with high-energy astrophysics: namely, X-ray, gamma-ray, and cosmic-ray astronomy. Since Earth's atmosphere absorbs most of the X-rays coming from celestial phenomena, astronomers must use high-altitude balloon platforms, sounding rockets, or orbiting spacecraft to study these interesting emissions, which are usually associated with very energetic, violent processes occurring in the universe. X-ray emissions carry detailed information about the temperature, density, age, and other physical conditions of the celestial objects that have produced them. The observation of X-ray emissions has been very valuable in the study of high-energy events, such as mass transfer in binary star systems, the interaction of supernova remnants with interstellar gas, and the functioning of quasars.

The first solar X-ray measurements were accomplished by rocket-borne instruments in 1949. Some 13 years later, Riccardo Giacconi (b. 1931) and Bruno B. Rossi (1905–93), along with their colleagues, detected the first nonsolar source of cosmic X-rays, called Scorpius X-1. They made the important (though unanticipated) discovery in June 1962, as a result of a sounding rocket flight. This event is often considered the start of X-ray astronomy. During the next eight years, instruments launched on rockets and balloons detected several dozen bright X-ray sources in the Milky Way Galaxy and a few sources in other galaxies. The excitement in X-ray astronomy grew when scientific spacecraft became available. Satellites allowed scientists to place complex instruments above Earth's atmosphere for extended periods of time. As a result, over 35 years, orbiting observatories have provided a greatly improved understanding of the energetic, often violent, phenomena that are now associated with X-ray emissions from astrophysical objects.

In December 1970, NASA launched *Explorer 42,* or the *Small Astronomical Satellite-1*, the first spacecraft devoted entirely to X-ray astronomy. This satellite, renamed *Uhuru* (the Swahili word for "freedom"), was lifted into Earth's orbit from San Marco, a rocket-launch platform off the coast of Kenya on the east coast of Africa. Successfully functioning until April 1973, the scientific spacecraft performed the first survey of the X-ray sky. In addition to detecting more than 300 X-ray sources, *Uhuru* also provided data about X-ray binaries and diffuse X-ray emission from galactic clusters. Since then, increasingly more sophisticated spacecraft has performed the observation of celestial X-ray emissions.

For example, in November 1978, NASA successfully launched the second *High-Energy Astronomy Observatory (HEAO-2)*—also called the *Einstein Observatory* in honor of physicist Albert Einstein. This massive 6,900-pound (3,130-kg) satellite contained a large grazing-incidence X-ray telescope that provided the first comprehensive images of the X-ray sky. Until then scientists studied cosmic X-ray sources mostly by determining their positions, measuring their X-ray spectra, and monitoring changes in their X-ray brightness over time. With *HEAO-2,* it became possible to routinely produce images of cosmic X-ray sources rather than simply locate their positions. This breakthrough in observation was made possible by the grazing-incidence X-ray telescope.

An X-ray is a very energetic packet (photon) of electromagnetic energy that cannot be reflected or refracted by glass mirrors and lens the way photons of visible light are focused in traditional optical telescopes. However, if an X-ray arrives almost parallel to a surface—that is, if the X-ray arrives at a grazing incidence—then the energetic photon can actually be reflected in a useful manner. The way the incident X-ray is reflected by this special surface depends on the atomic structure of the material and on the wavelength (energy level) of the X-ray. Using this grazing-incidence technique, scientists can arrange special materials to help "focus" incident X-rays (over a limited range of energies) onto an array of detection instruments.

The *Einstein Observatory (HEAO-2)* was the first such imaging X-ray telescope to be deployed in Earth's orbit. Its scientific objectives were to locate accurately and examine X-ray sources over the 0.2 to 4.0-keV energy range, to perform high-spectral-sensitivity spectroscopy, and to perform high-sensitivity measurements of transient X-ray sources. Operating successfully until 1981, the observatory provided astronomers with X-ray images of such extended optical objects as supernova remnants, normal galaxies, clusters of galaxies, and active galactic nuclei. Among the *Einstein Observatory*'s most unexpected discoveries was the finding that all stars, from the coolest to the very hottest, emit significant amounts of X-rays.

Thousands of cosmic X-ray sources became known due to observations by NASA's *Einstein Observatory* and the *European X-ray Observatory* satellite, launched by the European Space Agency in May 1983. As a result of these important discoveries in X-ray astronomy, astronomers now recognize that a significant fraction of the radiation emitted by virtually every type of interesting astrophysical object emerges as X-rays.

In June 1990, the joint German-U.S.-U.K. *Roentgen* satellite (ROSAT) was placed into orbit around Earth. This orbiting extreme-ultraviolet (EUV) and soft (low-energy) X-ray observatory was named in honor of Wilhelm Conrad Röentgen (1845–1923)—the German physicist who discovered X-rays. Some of the major scientific objectives of the very

X-RAY BURSTER

The X-ray burster is an X-ray source that radiates thousands of times more energy than the Sun, in short bursts that last only seconds. Astrophysicists suggest that a neutron star in a close binary star system accretes matter onto its surface from the companion until temperatures reach the level needed for hydrogen fusion to take place. The result is a sudden period of rapid nuclear burning and the subsequent release of an enormous quantity of energy.

The X-ray binary star system is the most often encountered type of luminous galactic X-ray source. It is a close binary star system in which material from a large, normal star flows (under gravitational forces) onto a compact stellar companion, such as a neutron star or a black hole (for the most luminous X-ray sources) or perhaps a white dwarf (for less luminous sources).

The gaseous matter pulled away from the companion star generally forms an accretion disk around the neutron star. Spiraling matter in the inner portions of the accretion disk becomes extremely hot, releasing a steady stream of X-rays. As the inward flowing gas builds up on the neutron star's surface, the temperature of the gas rises due to the pressure of

overlying material. Eventually, the temperature of the captured material reaches thermonuclear fusion conditions. The result is a rapid period of thermonuclear hydrogen burning—a phenomenon that releases an enormous amount of energy, characterized by a burst of X-rays. Hours then pass as fresh new accreted material accumulates on the neutron star's surface and conditions approach the production of another rapid X-ray burst.

Sometimes, not all of the in-falling material from the stellar companion makes it to the surface of the nearby neutron star. In such cases a portion of the in-falling gaseous material is shot out into interstellar space at enormously high speeds in narrow jets. For example, astronomers have observed an interesting celestial object called SS 433. This compact object expels the equivalent of one Earth mass of material each year in two narrow jets that travel in opposite directions roughly perpendicular to an accretion disk. When these jets (traveling at about 49,700 miles (80,000 km) per second—or about 25 percent of the speed of light)—interact with the interstellar medium, they emit radiation in the radio-frequency portion of the electromagnetic spectrum.

successful ROSAT mission were to study X-ray emission from stars of all spectral types, to detect and map X-ray emission from galactic supernova remnants, to perform studies of various active galaxy sources, and to perform a detailed EUV survey of the local interstellar medium. More recent orbiting X-ray observatories—such as NASA's *Chandra X-ray Observatory* and the European Space Agency's *XMM-Newton* Observatory—have been providing scientists higher spectral and angular resolution data and greater detection sensitivity. These sophisticated orbiting X-ray astronomy facilities are helping scientists understand many previously unresolved mysteries in high-energy astrophysics.

Formerly called the *Advanced X-ray Astrophysics Facility,* NASA launched the *Chandra X-ray Observatory* in July 1999 and named the space-based observatory after the Indian-American astrophysicist Subrahmanyan Chandrasekhar. This Earth-orbiting astrophysical facility has the capability to study some of the most interesting and puzzling X-ray sources in the universe, including emissions from active galactic nuclei, exploding stars, neutron stars, and matter falling into black holes. (Black holes are discussed in chapter 12.)

The European Space Agency's space observatory is also making significant contributions to contemporary X-ray astronomy. Launched on December 10, 1999, from Kourou, French Guiana, by an Ariane 5 rocket, *XMM-Newton* contains three very advanced X-ray telescopes. Previously called the *X-ray Multi-Mirror* satellite, *XMM-Newton* is the largest science satellite ever built in Europe—a very-high-sensitivity astrophysical instrument that can detect millions of X-ray sources.

Space-based observatories, such as the *Chandra X-ray Observatory* and *XMM-Newton,* as well as NASA's planned *Constellation X-ray Observatory,* provide modern astrophysicists with important X-ray emission data that they need more precisely understand stellar structure and evolution (including binary star systems, supernova remnants, pulsars, and black-hole candidates), large-scale galactic phenomena (including the interstellar medium itself and soft X-ray emissions of local galaxies), the nature of active galaxies (including the spectral characteristics and time variation of emissions from the central regions of such galaxies), and rich clusters of galaxies (including their associated X-ray emissions). A bit closer to home, X-ray emission data are also helping space scientists monitor violent and dangerous solar flares as they occur on our parent star, the Sun.

✧ NASA's *Chandra X-ray Observatory*

The *Chandra X-ray Observatory* (CXO) is one of NASA's four great orbiting astronomical observatories. This large and massive spacecraft was

In preparation for its launch, aerospace technicians at NASA's Kennedy Space Center prepare to attach and deploy a solar panel array on the *Chandra X-ray Observatory. (NASA/KSC)*

successfully launched on July 23, 1999, and deployed by the astronaut crew of the space shuttle *Columbia* during the STS-93 mission. During its development, NASA had called the scientific spacecraft the *Advanced X-ray Astrophysics Facility*. But after launch, NASA renamed this sophisticated X-ray observatory as the *Chandra X-ray Observatory* to honor the brilliant Indian-American astrophysicist and Nobel laureate Subrahmanyan Chandrasekhar (popularly known as Chandra). The Earth-orbiting astronomical facility studies some of the most interesting and puzzling X-ray sources in the universe, including emissions from active galactic nuclei, exploding stars, neutron stars, and matter falling into black holes.

Unlike the *Hubble Space Telescope* (*HST*), which operates in a circular orbit that is relatively close to Earth, NASA placed the *Chandra X-ray Observatory* in a highly elliptical (oval-shaped) orbit. To achieve its final operational orbit, the *CXO* was first carried into low Earth orbit by the space shuttle *Columbia* and then deployed from the orbiter's cargo bay by the astronaut crew into a 155-mile- (250-km-) altitude orbit above Earth. Next, two firings of an attached inertial upper stage rocket, followed by

SUBRAHMANYAN CHANDRASEKHAR (AKA: CHANDRA)

Subrahmanyan Chandrasekhar was the brilliant Indian–American astrophysicist who made important contributions to the theory of stellar evolution—especially the role of white dwarf as the last stage of evolution of many stars that are about the mass of the Sun. He shared the 1983 Nobel Prize in physics with William Alfred Fowler (1911–95) for his theoretical studies of the physical processes important to the structure and evolution of stars.

Known to the world as "Chandra" (which means "luminous" or "moon" in Sanskrit), Chandrasekhar was widely recognized as one of the foremost astrophysicists of the 20th century. He was born in Lahore, India, on October 19, 1910. Chandrasekhar trained as a physicist at Presidency College, Madras, India, and then became a Ph.D. student under Sir Ralph Howard Fowler (1889–1944) at Cambridge University in England. Early in his career, he demonstrated that there was an upper limit (now called the Chandrasekhar limit) to the mass of a white dwarf star.

In 1937 Chandrasekhar immigrated from India to the United States, where he joined the faculty of the University of Chicago. He and his wife (Lalitha Doraiswamy) became American citizens in 1953. Chandrasekhar was a popular professor at the University of Chicago, and his research interests extended to nearly all branches of theoretical astrophysics. He died in Chicago on August 21, 1995.

On July 23, 1999, NASA launched the *Advanced X-ray Astrophysics Facility* as part of the STS-93 mission of the space shuttle. Upon successful orbital deployment, NASA renamed this astronomical satellite the *Chandra X-ray Observatory* in his honor.

several firings of its own onboard propulsion system (after separation from the IUS rocket), placed the observatory into its highly elliptical working orbit around Earth. At its closest approach to Earth, the *CXO* has an altitude of about 5,965 miles (9,600 km). At its farthest orbital distance from Earth (about 87,000 miles [140,000 km]), the observatory travels almost one-third of the way to the Moon. The *CXO*'s working orbit is characterized by a period of 64.2 hours, an inclination of 28.5 degrees, and an eccentricity of 0.7984.

The observatory's highly elliptical orbit carries it far outside the trapped radiation belts that surround Earth. Exposure to the charged particle radiation found within the belts could easily upset the observatory's sensitive scientific instruments and provide faulty readings. So, NASA scientists use this special elliptical orbit to keep the *Chandra X-ray Observatory* outside the troublesome radiation belts long enough to take 55 hours of uninterrupted observations during each orbit. Of course, during the periods of interference from Earth's radiation belts (about nine hours each orbit), scientists do not attempt to make X-ray observations.

The *Chandra X-ray Observatory* has three major elements: the spacecraft system, the telescope system, and the science instruments. The spacecraft module contains computers, communications antennae, and data recorders to transmit and receive information between the observatory and ground stations. These onboard computers and sensors, along with human assistance from *CXO*'s ground-based control center, command and control the space vehicle and monitor its health throughout the operational lifetime (projected as a minimum of five years). The spacecraft module also has an onboard rocket-propulsion system to move and aim the entire observatory, an aspect camera that tells the observatory its position relative to the stars, and a Sun shade and sensor arrangement that protect sensitive components from excessive sunlight. Solar arrays provide electrical power to the spacecraft and also charge three nickel-hydrogen batteries, which serve as a backup electrical power supply.

At the heart of the telescope system is the high-resolution mirror assembly (HRMA). Since high-energy X-rays would penetrate a normal mirror, aerospace engineers created special cylindrical mirrors for the *Chandra X-ray Observatory*. The X-ray telescope consists of four nested paraboloid-hyperboloid X-ray mirror pairs, arranged in concentric cylinders within the cone. Basically, the HRMA is an assembly of tubes within tubes. Incoming X-rays graze off the highly polished mirror surfaces and are funneled to the instrument section for detection and analysis. The mirrors are slightly angled so that X-rays from sources in deep space will graze off their surfaces, much like a stone skips on a pond or lake. The function of HRMA is to accurately focus cosmic-source X-rays onto the imaging

A computer-enhanced rendering of the *Chandra X-ray Observatory,* in orbit with its solar panels fully deployed *(NASA/MSFC)*

instruments, which are located at the other end of the 32.8-foot- (10-m-) long telescope.

The *CXO*'s X-ray mirrors are the largest of their kind and the smoothest ever manufactured. If the surface of the state of Colorado were as relatively smooth, Pike's Peak (elevation 14,110 feet [4,300 m] above sea level) would be less than one inch (2.54 cm) tall. The largest of the eight mirrors is almost 3.9 feet (1.20 m) in diameter and 3.0 feet (0.91 m) long. The HRMA is contained in the cylindrical telescope portion of the observatory. The entire length of the telescope is covered with reflective multilayer insulation that assists heating elements inside the unit to maintain a constant internal temperature. By maintaining a precise internal temperature, the mirrors within the telescope are not subjected to expansion or contraction—thereby ensuring greater accuracy in scientific observations at X-ray portions of the electromagnetic spectrum. With its combination of large mirror area, accurate alignment, and efficient X-ray detectors, the *Chandra X-ray Observatory* has eight times greater resolution and is 20 to 50 times more sensitive than any previous X-ray telescope flown in space.

Within the instrument section of the observatory, two instruments at the narrow end of telescope structure collect X-rays and study them in various ways. Each of the *CXO*'s instruments can serves as an imager or a spectrometer. The high-resolution camera (HRC) records X-ray images, providing astrophysicists and astronomers an important look at violent, high-temperature celestial phenomena, such as the death of stars and colliding galaxies.

The HRC consists of two clusters of 69 million tiny, lead-oxide glass tubes. These lead-oxide tubes are only 0.05 inch (0.127 cm) long and just one-eighth the thickness of a human hair. When an X-ray strikes the tubes, electrons are released. As these charged particles are accelerated down the tubes by an applied high voltage, they cause an avalanche that involves perhaps 30 million more electrons in the process. A grid of electrically charged wires at the end of the tube bundle detects this flood of particles and allows the position of the original X-ray to be precisely determined. The high-resolution camera complements the *CXO*'s advanced charge-coupled device (CCD) imaging spectrometer (ACIS).

The observatory's imaging spectrometer is also located at the narrow end of the observatory. This detector is capable of recording not only the position but also the energy level (or the "color") of the incoming X-rays from cosmic sources. The imaging spectrometer consists of 10 CCD arrays. These detectors are similar to those found in home-video recorders and digital cameras but are designed to detect X-rays (as opposed to photons of visible light). Commands from the ground allow astronomers to select which of the detectors to use. The imaging spectrometer can distinguish up to 50 different energies within the range the observatory operates—that is, it can detect X-rays ranging from 0.09 to 10.0 kilo electron volts (keV).

In order to gain even more energy information, two screen-like instruments, called diffraction gratings, can be inserted into the path of the X-rays between the telescope and the detectors. The gratings change the path of the incident X-ray, and the *CXO*'s X-ray cameras record its energy level ("color") and position. How much the path of an X-ray photon changes depends upon on its initial energy level. One grating (called high-energy transmission grating [*HETG*]) concentrates on X-rays of higher and medium energies and uses the imaging spectrometer as a detector. The other grating (called the low-energy transmission grating [*LETG*]) disperses low-energy X-rays, and scientists use it in conjunction with the HRC.

By studying the X-ray spectra collected by the *CXO* and then recognizing the characteristic X-ray signatures of known elements, scientists can determine the composition of X-ray-producing objects and how these X-rays are produced. The principal mission objectives of the *Chandra X-ray Observatory* are to determine the nature of celestial objects from normal stars to quasars, to understand the nature of physical processes that take place in and between astronomical objects, and to help scientists study the history and evolution of the universe. In particular, the spacecraft is making observations of the cosmic X-rays from high-energy regions of the universe, such as supernova remnants, X-ray pulsars, black holes, neutron stars, and hot galactic clusters.

NASA's Marshall Space Flight Center has overall responsibility for the CXO mission. Under an agreement with NASA, the Smithsonian Astrophysical Observatory controls science and flight operations of the observatory from Cambridge, Massachusetts. The Smithsonian manages two electronically linked facilities: the operations control center and the science center.

The *CXO* operations control center is responsible for directing the observatory's mission as it orbits Earth. A control-center team interacts with the *CXO* three times a day, receiving science and housekeeping information from its recorders. The control-center team also sends new instructions to the observatory as needed. Finally, the control-center team transmits scientific information from the X-ray observatory to the *CXO* science center.

The science center is an important resource for scientists who wish to investigate X-ray–emitting objects such as quasars and colliding galaxies. The science-center team provides user support to qualified researchers throughout the scientific community. This support primarily involves the processing and archiving of the *Chandra X-ray Observatory*'s science data.

✧ NASA's *Rossi X-ray Timing Explorer*

The Rossi X-ray Timing Explorer is a NASA astrophysics mission designed to study the temporal and broadband spectral phenomena associated with stellar and galactic systems containing compact X-ray–emitting objects. The X-ray energy range observed by this spacecraft extended from two to 200 kiloelectron volts (keV), and the timescales monitored varied from microseconds to years.

The 7,040-pound (3,200-kg) spacecraft carried three special instruments to measure X-ray emissions from deep space—the proportional counter array (PCA), the high-energy X-ray timing experiment (HEXTE), and the all-sky monitor (ASM). The PCA and the HEXTE worked together to form a large X-ray observatory that was sensitive to X-rays from two to 200 keV. The ASM instrument observed the long-term behavior of X-ray sources and also served as a sentinel, monitoring the sky and enabling the spacecraft to swing rapidly to observe targets of opportunity with its other two instruments. Working together, these instruments gathered data about interesting X-ray emissions from the vicinity of black holes and from neutron stars and white dwarfs, along with telltale energetic electromagnetic radiation from exploding stars and active galactic nuclei.

NASA successfully launched this spacecraft into a 360-mile- (580-km-) altitude circular orbit around Earth on December 30, 1995, with an expendable Delta II rocket from Cape Canaveral, Florida. Following launch NASA renamed the spacecraft the *Rossi X-ray Timing Explorer* in honor of Professor Bruno B. Rossi, the distinguished Italian-American physicist who helped pioneer the field of X-ray astronomy. This spacecraft is also referred to as *Explorer 69* and the *NASA X-ray Timing Explorer*.

✧ NASA's Planned *Constellation X-ray Observatory*

The *Constellation X-ray Observatory* (also called simply *Constellation-X*) is a future NASA scientific mission that involves an array of X-ray satellites working together in tight orbit to improve by a hundredfold how astronomers and astrophysicists observe the universe in the X-ray portion of the electromagnetic spectrum. The current plan calls for four satellites operating in unison to generate the observing power of one giant space-based X-ray telescope.

The *Constellation-X* satellites will house high-resolution X-ray spectroscopy telescopes that collect high-energy X-rays produced by cataclysmic cosmic events and then interpret those event-related X-rays as spectra. Scientists regard X-ray spectra as the fingerprints of the chemicals producing the X-rays. When observations begin (about 2010), data from *Constellation-X* will help scientists resolve many pressing issues that currently challenge their understanding of the laws of physics. For example, *Constellation-X* observations of iron spectra in the vicinity of suspected massive black holes will help astrophysicists test Albert Einstein's general theory of relativity in an environment of extreme gravity. *Constellation-X* will also help scientists determine how black holes evolve and generate energy, thereby providing important information about the total energy content of the universe. With data from this observatory, scientists will also be able to investigate galaxy formation, the evolution of the universe on large scales, the nature of dark matter, and how the universe recycles its matter and energy—for example, how heavier elements from the cores of exploding stars eventually form planets and comets and how the gas from old stars helps make new ones.

Like all X-ray telescopes, *Constellation-X* must operate in outer space because X-ray photons from cosmic phenomena do not penetrate very far into Earth's atmosphere. The scientists and engineers who designed *Constellation-X* wanted to create an X-ray observatory capable of collecting as much X-ray "light" as possible—imitating to the greatest extent possible the way giant, ground-based optical telescopes, such as the Keck Telescope, use their large optics to gather as much visible light as possible from distant celestial objects. These demanding requirements led NASA personnel to select a rather unique multi-satellite design for *Constellation-X*. The four satellites will be of identical, low-mass design, allowing each spacecraft to be launched individually or possibly in pairs. The construction of four identical spacecraft reduces the overall cost of the mission and also avoids the risk of complete mission failure should a single launch abort occur. Once successfully co-orbited, the combined capability of

An artist's rendering of NASA's planned Constellation-X mission—a team of powerful X-ray telescopes that orbit close to each other and work in unison to simultaneously observe the same distant objects. By combining their data, the constellation of satellites becomes 100 times more powerful than any previous single X-ray telescope. *(NASA)*

Constellation-X's four X-ray telescopes will provide a level of sensitivity that is 100 times greater than any past or current X-ray satellite mission.

Essentially, scientists using *Constellation-X* will be able to collect more data in an hour than they can now collect in days or weeks with current space-based X-ray telescopes. Of special importance to astronomers and astrophysicists is the fact that *Constellation-X* will also allow them to discover and analyze thousands of faint X-ray emitting sources, not just the bright sources available today. NASA's *Constellation X-ray Observatory* promises to stimulate a revolution in X-ray astronomy and a much deeper understanding of energetic phenomena taking place throughout the universe.

Infrared Astronomy and the *Spitzer Space Telescope*

Infrared (IR) astronomy is the branch of modern astronomy that studies and analyzes IR radiation from celestial objects. Most celestial objects emit some quantity of IR radiation. However, when a star is not quite hot enough to shine in the visible portion of the electromagnetic spectrum, it emits the bulk of its energy in the infrared. IR astronomy, consequently, involves the study of relatively cool celestial objects, such as interstellar clouds of dust and gas (typically about 100 K) and stars with surface temperatures below temperatures of about 6,000 K.

Many interstellar dust and gas molecules emit characteristic IR signatures that astronomers use to study chemical processes occurring in interstellar space. This same interstellar dust also prevents astronomers from viewing visible light coming from the center of the Milky Way Galaxy. However, IR radiation from the galactic nucleus is absorbed not as severely as radiation in the visible portion of the electromagnetic spectrum, and IR astronomy enables scientists to study the dense core of the Milky Way.

IR astronomy also allows astrophysicists to observe stars as they are being formed (they call these objects protostars) in giant clouds of dust and gas (called nebulae), long before their thermonuclear furnaces have ignited and they have "turned on" their visible emission.

Unfortunately, water and carbon dioxide in Earth's atmosphere absorb most of the interesting IR radiation arriving from celestial objects. Earth-based astronomers can use only a few narrow IR spectral bands or windows in observing the universe, and even these IR windows are distorted by "sky noise" (undesirable IR radiation from atmospheric molecules). With the arrival of the space age, however, astronomers have placed sophisticated IR telescopes (such as the *Infrared Astronomical Satellite* [*IRAS*]) in space, above the limiting and disturbing effects of Earth's atmosphere, and have

An artist's rendering of NASA's *Spitzer Space Telescope* in orbit against an infrared (100-micrometer wavelength) sky *(NASA/JPL-Caltech)*

produced comprehensive catalogs and maps of significant IR sources in the observable universe.

The *Infrared Astronomical Satellite*, which was launched in January 1983, was the first extensive scientific effort to explore the universe in the IR portion of the electromagnetic spectrum. IRAS was an international effort involving the United States, the United Kingdom, and the Netherlands. By the time IRAS ceased operations in November 1983, this successful space-based infrared telescope had completed the first all-sky survey in a wide range of IR wavelengths with a sensitivity that was 100 to 1,000 times greater than any previous telescope. The impressive scientific results achieved by the 10-month IRAS mission demonstrated the promise of IR astronomy performed in space.

✧ The *Spitzer Space Telescope*

The *Spitzer Space Telescope* (*SST*) is the final spacecraft in NASA's Great Observatories Program—a family of four orbiting observatories each studying the universe in a different portion of the electromagnetic spectrum. The *Spitzer Space Telescope*—previously called the *Space Infrared*

Telescope Facility—consists of a 2.8-foot- (0.85-m-) diameter telescope and three cryogenically cooled science instruments. NASA renamed this space-based infrared telescope to honor the American astronomer Lyman Spitzer Jr. (See chapter 1 for a biographical sketch about Lyman Spitzer Jr.)

The *SST* represents the most powerful and sensitive infrared telescope ever launched. The orbiting facility obtains images and spectra of celestial objects at infrared radiation wavelengths between three and 180 micrometers—an important spectral region of observation mostly unavailable to ground-based telescopes because of the blocking influence of Earth's atmosphere. Following a successful launch on August 25, 2003, from Cape Canaveral Air Force Station by an expendable Delta rocket, the 2,094-pound- (950-kg-) mass observatory traveled to an Earth-trailing heliocentric orbit. Engineers and mission planners selected this operating orbit to allow the telescope instruments to cool rapidly with a minimum expenditure of onboard cryogen. With a planned mission lifetime in excess of five years, the *SST* has taken its place alongside NASA's other great orbiting astronomical observatories and is now collecting high-resolution infrared data that help scientists better understand how galaxies, stars, and planets form and develop.

One major engineering breakthrough with the *Spitzer Space Telescope* was the clever choice of orbit. Instead of orbiting Earth itself, the observatory now trails behind Earth as the planet orbits the Sun. The spacecraft drifts slowly away from Earth into deep space, circling the Sun at a distance of one astronomical unit, which is the mean Earth-Sun distance of 93 million miles (or 150 million km). The telescope is drifting away from Earth at about one-10th of one astronomical unit per year. This unique orbital trajectory keeps the observatory away from much of Earth's heat, which can reach −10°F (−23°C), or 250 K, for satellites and spacecraft in more conventional near-Earth orbits. The *Spitzer Space Telescope* operates in a more benign thermal environment for infrared telescopes—about −397°F (−238°C), or 35 K. With this innovative design approach, engineers have allowed natural radiation heat-transfer processes to the frigid deep-space environment to assist in keeping the observatory properly chilled.

Furthermore, the *SST*'s Earth-trailing orbit protects the observatory from Earth's radiation belts. This significantly reduces the harmful effects of ionizing radiation on the observatory's extremely sensitive infrared radiation detectors.

The infrared energy collected by the observatory is being examined and recorded by three main science instruments: an infrared array camera, an infrared spectrograph, and a multiband imaging photometer. The infrared array camera supports imaging at near- and mid-infrared wavelengths. Astronomers use this general-purpose camera for a wide variety of science research programs. The infrared spectrograph allows for both high- and low-resolution spectroscopy at mid-infrared wavelengths. Similar to an optical spectrometer, the infrared spectrograph spreads incoming infrared radiation into its constituent wavelengths. Scientists then scrutinize these infrared spectra for emission and absorption lines—the telltale fingerprints of atoms and molecules. The *Spitzer Space Telescope*'s spectrometer has no moving parts. Finally, the multiband imaging photometer provides imaging and limited spectroscopic data at far-infrared wavelengths. The only moving part in the imaging photometer is a scan mirror for efficiently mapping large areas of the sky.

The *Spitzer Space Telescope*'s powerful combination of highly sensitive detectors and its long lifetime allows astronomers to view objects and phenomena that have managed to elude them when they used other observing instruments and astronomical methods. Because of its unique and efficient thermal design, the spacecraft carries only 95 gallons (360 L) of expendable liquid helium cryogen to cool its sensitive infrared instruments. Cryogen depletion has severely limited the useful lifetime of previous infrared telescopes deployed in space. NASA mission planners estimate that *Spitzer*'s cryogen supply is sufficient to provide cooling for the infrared observatory's instrument for about five years of operation. The observatory uses the vapor from the boil-off of its cryogen to cool the infrared telescope assembly down to its optimal operating temperature of –450°F (–268°C), or 5.5 K.

The vast majority of the telescope's observing time is available to the general scientific community through peer-reviewed proposals. The observatory's final design was driven by the goal of making major scientific contributions in the following research areas: formation of planets and stars (including planetary debris disks and brown dwarf surveys), origin of energetic galaxies and quasars, distribution of matter and galaxies (including galactic halos and missing mass issue), and the formation and evolution of galaxies (including protogalaxies).

Protoplanetary and planetary debris disks are flattened disks of dust that surround many stars. Protoplanetary disks include large amounts of gas and are presumed to be planetary systems in the making. Planetary debris disks have most of their gas depleted and represent a more mature planetary system. The remaining dust disk may include gaps indicative of fledgling planetary bodies. By observing dust disks around stars at various ages, the *Spitzer Space Telescope* can trace the dynamics and chemical history of evolving planetary systems and provide statistical evidence of planetary system formation. (Chapter 11 discusses the contemporary search for extrasolar planets.)

Brown dwarfs are curious infrared objects that do not possess enough mass to gravitationally contract to the point of igniting nuclear-fusion reactions in their cores—in the same way that powers true stars. Astronomers consequently call brown dwarfs failed stars. Brown dwarfs are larger and warmer than the planets found in the solar system. At one point, brown dwarfs were considered just a theory, but astronomers have now begun to detect these long-sought objects. High-resolution infrared telescopes, such as the *Spitzer Space Telescope,* play a major role in the contemporary search. If brown dwarfs prove to be numerous enough, then they may represent an appreciable fraction of the of the elusive dark matter or missing mass issue that is now challenging scientists. (The search for dark matter and brown dwarfs is discussed in chapter 11.)

Many galaxies emit more radiation in the infrared portion of the spectrum than in all the other wavelength regions combines. These ultraluminous infrared galaxies could be powered by intense bursts of star formation, stimulated by colliding galaxies or by central black holes. The *Spitzer Space Telescope* can trace the origins and evolution of ultraluminous infrared galaxies out to cosmological distances (that is, billions of light-years away).

The *Spitzer Space Telescope* is examining galaxies at the cosmic fringe. These objects are so remote that the radiation they once emitted has taken billions of years to reach Earth. A consequence of an expanding universe, these faraway galaxies are receding from Earth so rapidly that most of their optical and ultraviolet light has red-shifted (Doppler effect) into the infrared portion of the spectrum. The *Spitzer Space Telescope* is examining some of these first stars and galaxies, in order to provide scientists with new clues into the character of the infant universe.

Apart from these important research areas in astronomy, the *Spitzer Space Telescope*'s near-infrared instrument can peer through obscuring dust, which cocoons newborn

A composite infrared image (at four wavelengths: 3.6 micrometers [μm], 4.5 μm, 5.8 μm, and 8.0 μm) taken by NASA's *Spitzer Space Telescope* on April 21, 2004. It shows a previously unknown globular cluster that lies in the dusty galactic plane about 9,000 light-years away. The inset (upper right) is a visible light image of the same region in the constellation Aquila and shows only a dark patch of sky where this large globular cluster is located. To help the reader appreciate the important role space-based infrared telescopes play in modern astronomy. As viewed from Earth, this new cluster's apparent size is comparable to a grain of rice held at arm's length. *(NASA/JPL-Caltech/H. Kobulnicky [University of Wyoming]; inset: NASA and the Digitized Sky Survey [DSS], California Institute of Technology)*

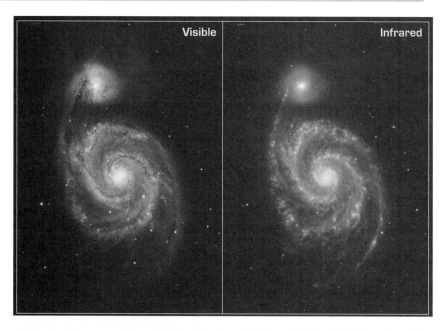

A side-by-side comparison of the mysterious Whirlpool Galaxy (M51) as viewed in the visible portion of the spectrum (left) and in the infrared portion of the spectrum (right). The visible image comes from the Kitt Peak National Observatory, while NASA's *Spitzer Space Telescope* collected the composite infrared image in May 2004. The light seen in the two images originates from very different sources. The infrared image shows astronomers the distributions of dust and stars between the Whirlpool Galaxy's spiral and its faint companion, called NGC 5195. The Whirlpool Galaxy is about 37 million light-years from Earth. *(NASA/JPL-Caltech/R. Kennicutt [University of Arizona]/DSS)*

stars, both in the nearby universe and also in the center of the Milky Way Galaxy. As in the past history of astronomy, whenever there is a giant leap in observational capability, like the kind that the *Spitzer Space Telescope* provides in infrared astronomy, there will also be a large number of astronomical surprises and serendipitous discoveries of unanticipated phenomena.

The *Spitzer Space Telescope* also captured a spectacular set of infrared images of the Whirlpool Galaxy (M51) that revealed strange structures bridging gaps between the dust-rich spiral arms and tracing the dust, gas, and stellar populations in both the bright spiral galaxy and its companion. The *Spitzer* image is a four-false-color composite of invisible light showing emissions from wavelengths of 3.6 micrometers (μm), 4.5 μm, 5.8 μm, and 8.0 μm. (Astronomers assign false colors to infrared wavelengths in *Spitzer Space Telescope* images to help them form more easily understandable pictures of normally invisible-to-the-human-eye infrared radiation data.) The visible image on page 136 comes from the Kitt Peak National Observatory.

Of special note here is the fact that the "light" (radiation) seen in these two (side-by-side) images originates from very different sources. At shorter wavelengths (in the visible bands and in the infrared from 3.6 to 4.5 micrometers wavelength), the light comes mainly from stars. This starlight fades at longer wavelengths (5.8 to 8.0 micrometers), where scientists see the glow from clouds of interstellar dust. This dust consists mainly of a variety of carbon-based organic molecules known collectively as polycyclic aromatic hydrocarbons. Wherever these compounds are found, there will also be dust granules and gas, which provide a reservoir of raw materials for future stars.

Astronomers think the spectacular whirlpool structure and star formation in M51 are being triggered by an ongoing collision with its (smaller) companion galaxy, called NGC 5195. Understanding the impact on star formation by the interaction of galaxies is one of the goals of a series of *Spitzer Space Telescope* observations known as the Spitzer Infrared Nearby Galaxy Survey. The Whirlpool Galaxy (M51) is a favorite with astronomers and lies about 37 million light-years away from Earth.

✧ NASA's *James Webb Space Telescope*

NASA's *James Webb Space Telescope* (*JWST*), previously called the *Next Generation Space Telescope* (*NGST*), is scheduled for launch in about 2010. An expendable launch vehicle will send the spacecraft on a three-month journey to its operational location, about 0.93 million miles (1.5 million km) from Earth in orbit around the second Lagrange libration point (or L2) of the Earth-Sun system. This distant location will provide an important advantage for the spacecraft's infrared radiation (IR) imaging and spectroscopy instruments. At that location, a single shield on just one side of the observatory will protect the sensitive telescope from unwanted thermal radiation from both the Earth and the Sun. As a result, the spacecraft's infrared sensors will be able to function at a temperature of −370°F (−223°C), or 50 K, without the need for complicated refrigeration equipment.

The telescope is being designed to detect electromagnetic radiation whose wavelength lies in the range from 0.6 to 28 micrometers but it will have an optimum performance in the one to five micrometer region. The *JWST* will be able to collect infrared signals from celestial objects that are much fainter than those now being studied with very large, ground-based infrared telescopes (such as the Keck Observatory) or the current generation of space-based infrared telescopes (such as the *Spitzer Space*

Telescope). With a primary mirror diameter of at least 19.7 feet (6 m), the *JWST* will make infrared measurements that are comparable in spatial resolution (that is, image sharpness) to images currently collected in the visible-light portion of the spectrum by the *Hubble Space Telescope.*

In September 2002, NASA officials renamed the *NGST* to honor James E. Webb—the civilian space agency's administrator from February 1961 to October 1968, during the development of the Apollo Project. Webb also initiated many other important space science programs within the fledgling agency. NASA is developing the *JWST* to observe the faint infrared signals from the first stars and galaxies in the universe. Data from the orbiting observatory should help scientists better respond to lingering fundamental questions about the universe's origin and ultimate fate. One important astrophysical mystery involves the nature and role of dark matter.

Ultraviolet Astronomy and the *Extreme Ultraviolet Explorer*

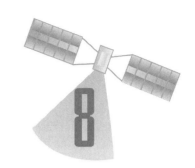

Ultraviolet (UV) astronomy is the branch of astronomy based on the ultraviolet (10 to 400 nanometer wavelength) portion of the electromagnetic spectrum. UV radiation is that portion of the electromagnetic spectrum which lies beyond visible (violet) light and is longer in wavelength than X-rays. Generally, astronomers and astrophysicists consider UV radiation as electromagnetic radiation with wavelengths between 4,000 angstroms (400 nm)—just past violet light in the visible spectrum—and about 100 angstroms (10 nm)—the extreme ultraviolet cutoff and the beginning of X-rays. Scientists sometimes find it useful to use the term *extreme ultraviolet radiation* for radiation in that region of the electromagnetic spectrum corresponding to wavelengths between 100 and 1,000 angstroms (10 and 100 nm).

Because of the strong absorption of UV radiation by Earth's atmosphere, UV astronomy is best performed using high-altitude balloons, rocket probes, and orbiting observatories. UV data gathered from spacecraft are extremely useful in investigating interstellar and intergalactic phenomena. Observations in the UV wavelengths have shown, for example, that the very-low-density material that can be found in the interstellar medium is quite similar throughout the Milky Way Galaxy but that its distribution is far from homogeneous. In fact, UV data have led some astrophysicists to postulate that low-density cavities, or bubbles, in interstellar space are caused by supernova explosions and are filled with gases that are much hotter than the surrounding interstellar medium.

UV data gathered from space-based observatories have revealed that some stars blow off material in irregular bursts and not in a steady flow as originally thought. Astrophysicists also find that UV data are of considerable use when they want to study many of the phenomena that occur in distant galaxies, including active galactic nuclei, which are often associated with massive black holes at the centers of galaxies.

✧ *International Ultraviolet Explorer*

The *International Ultraviolet Explorer* (*IUE*) was a highly successful scientific spacecraft launched in January 1978. Operated jointly by NASA and the European Space Agency (ESA), the *IUE* helped astronomers from around the world obtain access to the ultraviolet (UV) radiation of celestial objects in unique ways not available by other means.

The spacecraft contained a 1.5-foot- (0.45-m-) aperture telescope solely for spectroscopy in the wavelength range from 1,150 to 3,250 angstroms (115 to 325 nm). IUE data have helped support fundamental studies of comets and their evaporation rate when they approach the Sun and of the mechanisms driving the stellar winds that make many stars lose a significant fraction of their mass—before they die slowly as white dwarfs or suddenly in supernova explosions. This long-lived international spacecraft also assisted astrophysicists in their search to understand the ways by which black holes possibly power the turbulent and violent nuclei of active galaxies. Working nonstop since its launch, ESA mission controllers finally turned *IUE* off in September 1996. Of special note: *IUE* provided

SUPERNOVA 1987A

Supernova 1987A (SN 1987A) was a spectacular Type II supernova event observed in the Large Magellanic Cloud (LMC) starting on February 24, 1987. This was the first supernova bright enough to be seen by the naked eye (for observers in the Southern Hemisphere) since the supernova of 1604, which is also called Kepler's star.

Because this event took place relatively close to Earth, scientists from around the world used a variety of space-based, balloon-borne, and ground-based instruments to examine the explosive death of a red supergiant star. For example, almost a day (some 20 hours) before SN 1987A was detected optically, giant underground neutrino detectors in both Japan and the United States simultaneously recorded a brief (13-second) burst of neutrinos from the supernova. This successful data collection was a major milestone in neutrino astronomy and astrophysics because it marked the first time scientists have collected information about a particular celestial object beyond the solar system that does not involve radiation within the electromagnetic spectrum.

Balloon-borne and space-based gamma-ray detectors provided confirmation that heavier elements (such as cobalt) were also produced in this supernova. SN 1987A remained visible to the naked eye until the end of 1987. But the study of this interesting event did not end there. Today scientists are using sophisticated space-based observatories such as the *Chandra X-ray Observatory* to carefully analyze the expanding debris (supernova remnant) from this gigantic explosion.

space-based observations of the first naked-eye visible supernova event in 300 years—the 1987 supernova in the Large Magellanic Cloud (LMC), a nearby galaxy about 160,000 light-years away.

✧ Extreme Ultraviolet Explorer

The *Extreme Ultraviolet Explorer* (*EUVE*) was one of NASA's Explorer-class satellites launched from Cape Canaveral Air Force Station by a Delta II rocket on June 7, 1992. The scientific satellite traveled around Earth in an approximate 325-mile- (525-km-) altitude orbit, with a period of 94.8 minutes and an inclination of 28.4 degrees. It provided astronomers with a view of the relatively unexplored region of the electromagnetic spectrum—the extreme ultraviolet (EUV) region (i.e., 10 to 100 nanometers wavelength).

The science payload consisted of three grazing-incidence scanning telescopes and an EUV spectrometer/deep-survey instrument. The science payload was attached to NASA's Multi-Mission Modular spacecraft. For the first six months following its launch, the spacecraft performed a full-sky survey. The spacecraft also gathered important data about sources of EUV radiation within the "Local Bubble"—a hot, low-density region of the Milky Way Galaxy (including the Sun) that is the result of a supernova explosion some 100,000 years ago. Interesting EUV sources include white dwarf stars and binary star systems in which one star is siphoning material from the outer atmosphere of its companion.

NASA extended the EUVE mission twice, but by the year 2000, operational costs and scientific merit issues led to a decision to terminate spacecraft activities. Consequently, NASA commanded the satellite's transmitters off on January 2, 2001, and then formally ended satellite operations on January 31, 2001, by placing the spacecraft in a safehold configuration. On January 30, 2002, the *EUVE* spacecraft reentered Earth's atmosphere over central Egypt. This spacecraft was also known as *BERKSAT* and *Explorer 67*.

✧ Far Ultraviolet Spectroscopic Explorer

NASA's *Far Ultraviolet Spectroscopic Explorer* (*FUSE*) represented the next-generation, high-orbit, ultraviolet space observatory by examining the wavelength region of the electromagnetic spectrum ranging from about 900 to 1,200 angstroms (90 to 120 nm).

This scientific spacecraft was launched from Cape Canaveral on June 24, 1999, by a Delta II rocket. The ultraviolet astronomy satellite now travels around Earth in a near-circular 467-mile × 477-mile (752-km × 767-km) orbit at an inclination of 25 degrees.

The primary objective of *FUSE* is to use high-resolution spectroscopy at far-ultraviolet wavelengths to study the origin and evolution of the lightest elements (hydrogen and deuterium) created shortly after the big bang, and the forces and processes involved in the evolution of galaxies, stars and planetary systems. *FUSE* is a part of NASA's Origins Program. The spacecraft represents a joint U.S.-Canada-France scientific project. A previous mission, involving the *Copernicus* spacecraft, examined the universe in the far-ultraviolet region of the electromagnetic spectrum. However, *FUSE* is providing data collection with sensitivity some 10,000 times greater than the *Copernicus* spacecraft.

The *FUSE* satellite consists of two primary sections: the spacecraft and the science instrument. The spacecraft contains all of the elements necessary for powering and pointing the satellite, including the attitude-control system, the solar panels, and communications electronics and antennae. The observatory is approximately 25 feet (7.6 m) long with baffle fully deployed.

The *FUSE* science instrument consists of four co-aligned telescope mirrors (with approximately a 15.4-inch × 13.8-inch [39-cm × 35-cm] clear aperture). The light from the four optical channels is dispersed by four spherical, aberration-corrected holographic diffraction gratings and is recorded by two delay-line microchannel plate detectors. Two channels with SiC coatings cover the spectral range from 905 to 1100 angstroms (90.5 to 110 nm), and two channels with LiF coatings cover the spectral range from 1,000 to 1,195 angstroms (100 to 119.5 nm).

The *FUSE* observatory was designed for an operational lifetime of three years, although mission planners hope it may remain in operation for as long as 10 years. Through 2006, *FUSE* continued to operate in a satisfactory matter and provide valuable observational data, although a problem with the pointing system cause some difficulties in late 2001.

✧ NASA's *Galaxy Evolution Explorer*

On April 28, 2003, NASA launched the *Galaxy Evolution Explorer* (*GALEX*) from Cape Canaveral, Florida. The spacecraft is an orbiting space-based ultraviolet telescope with a mission to study the shape, brightness, size, and distance of galaxies across 10 billion years of cosmic history.

The 620-pound (280-kg), three-axis stabilized spacecraft is a cylinder measuring about three feet (1 m) in diameter and is 8.2 feet (2.5 m) high. Much of the spacecraft's flight software is derived from software previously developed for NASA's *Far Ultraviolet Spectroscopic Explorer*. All of the satellite's computing functions are performed by the command- and data-handling subsystem. Among the tasks managed by the computer are Sun avoidance; deployment of the solar arrays; precision determination and control of the satellite's orientation (attitude); thermal management; automated fault detection and correction; communication with the telescope instrument; and acquisition, storage, and transmission of science data. A 24-gigabit solid-state recorder stores engineering data from the satellite and science instrument and science data from the telescope.

At the heart of the *Galaxy Evolution Explorer* is a telescope designed to look out into space from Earth orbit. In certain ways, this telescope is like a much smaller version of the *Hubble Space Telescope* (*HST*). Besides being quite a bit smaller that *HST,* the *GALEX* telescope has fewer instruments to record the light gathered by its main mirror. In addition, it is optimized for one specialty, surveying galaxies in ultraviolet light.

Like the *Hubble Space Telescope*, the *GALEX* telescope is of a Cassegrain design, named after Frenchman Guillaume Cassegrain (ca. 1629–93), who invented it in 1672. In this design, light from distant objects in space enters the telescope and is reflected by a primary

An artist's rendering of NASA's *Galaxy Evolution Explorer*, which was launched from Cape Canaveral on April 28, 2003. The spacecraft's mission is to study the shape, brightness, size, and distance of galaxies across 10 billion years of cosmic history. *(NASA/JPL-Caltech)*

mirror at the telescope's rear. The light is then gathered onto a small (secondary) mirror suspended in the middle of the telescope near the front end. The light in turn reflects back toward the rear of the telescope, where it passes through a hole in the middle of the primary mirror. At the rear, behind the primary mirror, is the sensor that records the image.

The primary mirror in the *Galaxy Evolution Explorer*'s telescope is 19.7 inches (50 cm) in diameter. It is made of fused silica with a thin coating of aluminum. The aluminum must be protected by another coating of clear material, however, to keep it from oxidizing and degrading. In the *GALEX* telescope, the optics are coated with a material, called magnesium fluoride, which is transparent to ultraviolet light.

Scientists want to hunt star-birthing galaxies in both the near-ultraviolet and far-ultraviolet portions of the spectrum, so the *GALEX* telescope delivers the light it gathers to two separate detectors. It accomplishes this task in a rather ingenious way. Instead of swapping different detectors at different times, the telescope directs light through an unusually shaped lens that scientists call a dichroic beam-splitter. This lens is coated with many extremely thin layers of special materials that cause the far-ultraviolet light to reflect off the lens surface, while allowing the near-ultraviolet light to proceed unimpeded through the lens. The reflected light proceeds to the far-ultraviolet detector, while the light that passes through the lens proceeds to the near-ultraviolet detector. This allows both detectors to perform science observations at the same time. The near-ultraviolet detector responds to photons with wavelengths between 1,750 and 2,800 angstroms (175 and 280 nm), and the far-ultraviolet detector responds to photons with wavelengths between 1,350 and 1,740 angstroms (135 and 174 nm).

In addition to gathering basic ultraviolet images, scientists are also interested in analyzing the spectral signatures of light from distant galaxies to measure their redshift. Because the universe has been expanding since the big bang explosion, all of the galaxies in the universe are moving away from each other. Astronomers use the amount of redshift in a galaxy's spectrum to determine the galaxy's distance. Because these distances are enormous, the light from distant galaxies takes a significant fraction of the age of the universe to reach Earth. When they survey the universe, astronomers view galaxies of any age, back to the time when they first formed.

The *Galaxy Evolution Explorer*'s instrument is therefore equipped with a "grism" lens, mounted so that it can be rotated into the beam of light coming from the telescope. The grism gets its name from the fact that it is a prism with a grating on one surface. This lens breaks light into its various (color) wavelengths, which reveal telltale lines caused when light is absorbed or emitted by various elements. The grism on the *Galaxy Evolution Explorer* is made from calcium fluoride crystal, the first time such optical material has been flown in space.

Following its successful launch and placement into a nominal 416-mile- (670-km-) altitude circular

An ultraviolet image of the planetary nebula NGC 7293, which is also known as the Helix Nebula. It is the nearest example of what happens to a star, like the Sun, as it approaches the end of its life. When a Sun-like star runs out of (hydrogen) fuel for thermonuclear burning in its core, the dying star expels gas outward and evolves into a much hotter, smaller, and denser white dwarf star. NASA's *Galaxy Evolution Explorer* collected this image. *(NASA/JPL-Caltech/SSC)*

orbit around Earth at an inclination of 28.5 degrees, the *Galaxy Evolution Explorer* started blazing new trails in ultraviolet astronomy. The mission is providing the first-ever wide-area ultraviolet surveys of the sky and the first wide-area spectroscopic surveys. By observing large pieces of the sky all at once, the *Galaxy Evolution Explorer* is finding the most interesting and rare ultraviolet-emitting objects in the universe.

In addition to studying galaxies, the Galaxy Evolution Explorer mission is compiling a substantial archive of other objects of interest to astronomers. These include active galactic nuclei, often associated with massive black holes at the centers of galaxies; white dwarfs, old stars that have blown off their outer shells, leaving very hot cores that are bright in the ultraviolet; and quasars, thought to be associated with black holes and active galactic nuclei.

✧ Active Galaxies

A galaxy is a system of stars, gases, and dust bound together by mutual gravity. A typical galaxy has billions of stars, and some galaxies even have trillions of stars. Although galaxies come in many different shapes, the basic structure is the same: a dense core of stars called the galactic nucleus surrounded by other stars and gas. Normally, the core of an elliptical or disk galaxy is small, relatively faint, and composed of older, redder stars. However, in some galaxies, the core is intensely bright—shining with a power level equivalent to trillions of Sun-like stars and easily outshining the combined light from all of the rest of the stars in that galaxy. Astronomers call a galaxy that emits such tremendous amounts of energy an active galaxy (AG), and they call the center of an active galaxy the active galactic nucleus (AGN). Despite the fact that active galaxies are actually quite rare, because they are so bright, they can be observed at great distances—even across the entire visible universe.

Scientists currently believe that at the center of these bright galaxies lies a supermassive black hole—an incredibly massive object that contains the masses of millions or perhaps billions of stars the size of the Sun. As matter falls toward a supermassive black hole, the material forms an accretion disk—a flattened disk of gravitationally trapped material swirling around the black hole. Friction and magnetic forces inside the accretion disk heat the material to millions of kelvins, and it glows brightly almost all the way across the electromagnetic spectrum, from radio waves to X-rays. Although the Milky Way has a central supermassive black hole, it is not an active galaxy. For reasons that astrophysicists cannot currently explain, the black hole at the center of humans' home galaxy is inactive, or quiescent, as are most present-day galaxies.

Although the physics underlying the phenomenon is not well understood, scientists know that in some cases the accretion disk of an active galaxy focuses long jets of matter that streak away from the AGN at speeds near the speed of light. The jets are highly collimated (meaning they retain their narrow focus over vast distances) and are emitted in a direction perpendicular to the accretion disk. Eventually, these jets slow to a stop due to friction with gas well outside the galaxy, forming giant clouds of matter that radiate strongly at radio wavelength. In addition, surrounding the accretion disk is a large torus (donut-shaped cloud) of molecular material. When viewed from certain angles, this torus can obscure observations of the accretion disk surrounding the supermassive black hole.

There are many types of AGs. Initially, when astronomers were first studying active galaxies, they thought that the different types of active galaxies were fundamentally different celestial objects. Now many (but not all) astronomers and astrophysicists generally accept the unified model of AGs. This means that most or all active galaxies are actually just different versions of the same object. Many of the apparent differences between types of active galaxies are due to viewing the AG at different orientations with respect to the accretion disk, or due to observing the AG in different wavelength regions of the electromagnetic (EM) spectrum (such as the radio-frequency, visible, and X-ray portions of the EM spectrum).

Basically, the unified model of AGs suggests that the type of active galaxy astronomers see depends on the way they see it. If they see the accretion disk and gas torus edge on, they call the active galaxy a radio galaxy. The torus of cool gas and dust blocks most of the visible, ultraviolet, and X-ray radiation from the intensely hot inflowing material as it approaches the event horizon of the supermassive black hole or as it swirls nearby in the accretion disk. As a consequence, the most obvious observable features are the radio wave–emitting jets and giant lobes well outside the active galaxy.

If the disk is tipped slightly to astronomers' line of sight, they can see higher-energy (shorter-wavelength) electromagnetic radiation from the accretion disk inside the gas torus in addition to the lower-energy (longer-wavelength) electromagnetic radiation. Astronomers call this type of active galaxy a Seyfert galaxy—named after the American astronomer Carl Keenan Seyfert (1911–60), who first cataloged these galaxies in 1943. A Seyfert galaxy looks very much like a normal galaxy but with a very bright core (active galactic nucleus), and it may be giving off high-energy photons like X-rays.

If the active galaxy is very far away from Earth, astronomers may observe the core (AGN) as a starlike object even if the fainter surrounding galaxy is undetected. In this case, they call the active galaxy a quasar, which is scientific shorthand for quasi-stellar radio-source—so named because the first such objects detected appeared to be starlike through a telescope, but, unlike regular stars, emitted copious quantities of radio waves. The Dutch-American astronomer Maarten Schmidt (b. 1929) discovered the first quasar (called 3C 273) in 1963. This quasar is an active galaxy a very great distance away, and it is receding at more than 90 percent of the speed of light. Quasars are among the most distant, and therefore youngest, extragalactic objects astronomers can observe.

If the active galaxy is tipped 90 degrees with respect to observers on Earth, astronomers would be looking straight down a jet from the active galaxy. They call this type of object a blazar. The first blazar detected was a bl lac object. But, in the late 1920s, they mistakenly classified this type of extragalactic object as a variable star, because of its change in visual brightness. It was not until the 1970s that astronomers recognized the extragalactic nature of this interesting class of objects. More recently, using advanced space-based observatories, such as the *Compton Gamma Ray Observatory,* astronomers have detected very energetic gamma-ray emissions from blazers.

In summary, the basic components of an active galaxy are a supermassive black-hole core, an accretion disk surrounding this core, and a torus of gas and dust. In some, but not all cases, there is also a pair of highly focused jets of energy and matter. The type of active galaxy astronomers see depends upon the view angle at which they observe a particular active galaxy. The generally accepted unified model of active galaxies includes blazers, quasars, radio galaxies, and Seyfert galaxies.

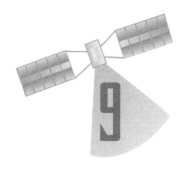

A Visit to the Nearest Star: Space-Based Solar Physics

An object of worship and awe by many ancient civilizations, the Sun has been well studied from the ground. Since the days of Galileo Galilei, scientists have known that the Sun rotates and is speckled with spots, though their knowledge of what these spots are and their 11-year cycle is relatively recent.

After early observations from sounding rockets following World War II, the scientific study of the Sun from space began in the early 1960s, when NASA launched the Orbiting Solar Observatory (OSO), which was a series of eight Earth-orbiting observatories sent into space between 1962 and 1971. Seven of these early scientific spacecraft were successful and studied the Sun at ultraviolet and X-ray wavelengths. The OSO spacecraft also photographed the million-degree solar corona, made X-ray observations of a solar flare, and enhanced scientific understanding of the Sun's atmosphere.

The Apollo Telescope Mount (ATM) was an innovative program for astronauts to observe the Sun from *Skylab*, the first American space station. The ATM was the most important scientific instrument on *Skylab*. Unhampered by telemetry limitations, the astronauts were able to bring back a large number of important images, including X-ray observations of solar flares, coronal holes, and the corona itself.

✧ The Sun: Humans' Parent Star

About eight light-minutes away from Earth, the Sun is the nearest star. It has a diameter of 0.86 million miles (1.39 million km) and a mass of 4.38×10^{30} pounds (1.99×10^{30} kg). The Sun is the massive, luminous celestial object around which all other bodies in the solar system revolve.

It provides the light and warmth upon which (almost) all terrestrial life depends.

The Sun's gravitational field determines the movement of the planets and other celestial bodies (e.g., comets). Astronomers classify the Sun as a main-sequence star of spectral type G2V. Like all main-sequence stars, the Sun derives its abundant energy output from thermonuclear fusion reactions involving the conversion of hydrogen to helium and heavier nuclei. Photons associated with these exothermic (energy-releasing) fusion reactions diffuse outward from the Sun's core, until they reach the convective envelope. Another by-product of the thermonuclear fusion reactions is a flux of neutrinos that freely escape from the Sun.

At the center of the Sun is the core, where energy is released in thermonuclear reactions. Surrounding the core are concentric shells, which form the radiative zone, the convective envelope (which occurs at approximately 0.8 of the Sun's

The solar eruption of June 10, 1973, as seen in this spectroheliogram obtained during NASA's Skylab mission. At the top of this image a great eruption can be observed extending more than one-third of a solar radius from the Sun's surface. In the picture, solar north is to the right and east is up. The wavelength scale (150 to 650 angstroms [15 to 65 nm]) increase is to the left. *(NASA)*

radius), the photosphere (the layer from which visible radiation emerges), the chromosphere, and, finally, the corona (the Sun's outer atmosphere). Energy is transported outward through the convective envelope by convective (mixing) motions that are organized into cells. The Sun's lower or inner atmosphere, the photosphere, is the region from which energy is radiated directly into space. Solar radiation approximates a Planck distribution (blackbody source) with an effective temperature of 5,800 K.

The chromosphere, which extends for a few thousand miles above the photosphere, has a maximum temperature of approximately 10,000 K. The corona, which extends several solar radii above the chromosphere, has temperatures of over 1 million K. These regions emit electromagnetic (EM) radiation in the ultraviolet (UV), extreme ultraviolet (EUV), and X-ray portions of the spectrum. This shorter-wavelength EM radiation, although representing a relatively small portion of the Sun's total energy output, still plays a dominant role in forming planetary ionospheres and in photochemistry reactions occurring in planetary atmospheres.

Since the Sun's outer atmosphere is heated, it expands into the surrounding interplanetary medium. This continuous outflow of plasma is called the solar wind. It consists of protons, electrons, and alpha particles, as well as small quantities of heavier ions. Typical particle velocities in the solar wind fall between 186 and 250 miles (300 and 400 km) per second, but these velocities may get as high as 620 miles (1,000 km) per second.

Although the total energy output of the Sun is remarkably steady, its surface displays many types of irregularities. These include sunspots, faculae, plages (bright areas), filaments, prominences, and flares. All are believed to be the result of interactions between ionized gases in the solar atmosphere and the Sun's magnetic field. Most solar activity follows the sunspot cycle. The number of sunspots varies, with a period of about 11 years. However, this approximately 11-year sunspot cycle is only one aspect of a more general 22-year solar cycle that corresponds to a reversal of the polarity patterns of the Sun's magnetic field.

Sunspots were originally observed by Galileo Galilei in 1610. They are less bright than the adjacent portions of the Sun's surface because they are not as hot. A typical sunspot temperature might be 4,500 K compared to the photosphere's temperature of about 5,800 K. Sunspots appear to be made up of gases boiling up from the Sun's interior. A small sunspot may be about the size of Earth, while larger ones could hold several hundred or even thousands of Earth-size planets. Extra-bright solar regions, called plages, often overlie sunspots. The number and size of sunspots appear to rise and fall through a fundamental 11-year cycle (or in an overall 22-year cycle, if polarity reversals in the Sun's magnetic field are considered). The greatest number occurs in years when the Sun's magnetic field is the most severely twisted (called sunspot maximum). Solar physicists think that sunspot migration causes the Sun's magnetic field to reverse its direction. It then takes another 22 years for the Sun's magnetic field to return to its original configuration.

A solar flare is the sudden release of tremendous energy and material from the Sun. A flare may last minutes or hours, and it usually occurs in complex magnetic regions near sunspots. Exactly how or why enormous amounts of energy are liberated in solar flares is still unknown, but scientists think the process is associated with electrical currents generated by changing magnetic fields. The maximum

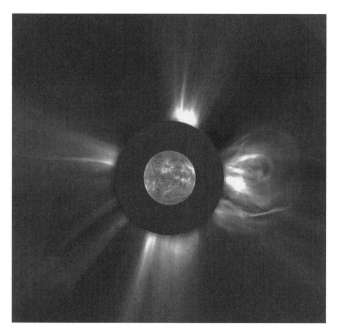

This *SOHO* spacecraft image shows the largest solar flare on record (up to that time). The gigantic explosion in the atmosphere of the Sun took place on April 2, 2001. *(NASA Goddard SOHO Project Office)*

number of solar flares appears to accompany the increased activity of the sunspot cycle. As a flare erupts, it discharges a large quantity of material outward from the Sun. This violent eruption also sends shock waves through the solar wind.

Data from space-based solar observatories have indicated that prominences (condensed streams of ionized hydrogen atoms) appear to spring from sunspots. Their looping shape suggests that these prominences are controlled by strong magnetic fields. About 100 times as dense as the solar corona, prominences can rise at speeds of hundreds of miles per second. Sometimes the upper end of a prominence curves back to the Sun's surface, forming a bridge of hot glowing gas hundreds of thousands of kilometers long. On other occasions, the material in the prominence jets out and becomes part of the solar wind.

High-energy particles are released into heliocentric space by solar events, including very-large solar flares called *anomalously large solar particle events* (ALSPEs). Because of their close association with infrequent large flares, these bursts of energetic particles are also relatively infrequent. However, solar flares, especially ALSPEs, represent a potential hazard to astronauts traveling in interplanetary space or working on the surface of the Moon or Mars.

✧ Skylab

Skylab was the first American space station. This Earth-orbiting facility was placed in orbit in 1973 by a two-stage configuration of the Saturn V expendable launch vehicle and then visited by three astronaut crews who worked on scientific experiments in space for a total of approximately 172 days, with the last crew spending 84 days in Earth orbit.

Skylab was composed of five major parts: the Apollo telescope mount (ATM); the multiple docking adapter (MDA); the airlock module; the instrument unit; and the orbital workshop, which included the living and working quarters. The ATM was a solar observatory that provided attitude control and experiment pointing for the rest of the cluster. The retrieval and installation of film used in the ATM was accomplished by the astronauts during extravehicular activity. The MDA served as a dock for the modified Apollo spacecraft that taxied the crews to and from the space station. The airlock module was located between the docking port (MDA) and the living and working quarters; it contained controls and instrumentation. The instrument unit, which was used only during launch and the initial phases of operation, provided guidance and sequencing functions for the initial deployment of the ATM, its solar arrays, and the like. The orbital workshop was a modified Saturn IV-B stage that had

American astronaut and solar physicist Edward G. Gibson is shown here working at the Apollo Telescope Mount console during NASA's Skylab 4 mission. The photograph was taken onboard the *Skylab* space station on December 5, 1973, during a live television broadcast from the Earth–orbiting facility down to Earth. *(NASA)*

been converted into a two-story-high space laboratory with living quarters for a crew of three. This orbital laboratory was capable of unmanned, in-orbit storage, reactivation, and reuse.

There were four launches in the Skylab Program from Complex 39 at the Kennedy Space Center. The first launch was on May 14, 1973. A two-stage Saturn V vehicle placed the unmanned 100-ton (90-metric ton) *Skylab* space station in an initial 270-mile- (435-km-) altitude orbit around Earth. As the rocket accelerated past an altitude of 25,000 feet (7,620 m), atmospheric drag began clawing at *Skylab*'s meteoroid/Sun shield. This cylindrical metal shield was designed to protect the orbital workshop from tiny particles and the Sun's scorching heat. Sixty-three seconds after launch, the shield ripped away from the spacecraft, trailing an aluminum strap that caught on one of the unopened solar wings. The shield became tethered to the laboratory while at the same time prying the opposite solar wing partly open. Minutes later, as the booster rocket staged, the partially deployed solar wing and meteoroid/Sun shield were flung into space. With the loss of the meteoroid/Sun shield, temperatures inside Skylab soared, rendering the space station uninhabitable and threatening the food, medicine, and film stored onboard. The ATM, the major piece of scientific equipment, did deploy properly, however, an action that included the successful unfolding of its four solar panels.

The countdown for the launch of the first Skylab crew was halted. NASA engineers worked quickly to devise a solar parasol to cover the workshop and to find a way to free the remaining stuck solar wing. On May 25, 1973, astronauts Charles (Pete) Conrad Jr., Joseph P. Kerwin, and Paul J. Weitz were launched by a Saturn 1B rocket toward *Skylab*. After repairing *Skylab*'s broken docking mechanism, which had refused to latch, the astronauts entered the space station and erected a mylar solar parasol through a space access hatch. It shaded part of the area where the protective meteoroid/Sun shield had been ripped away. Temperatures within the spacecraft immediately began dropping, and *Skylab* soon became habitable without space suits. But the many experiments on board demanded far more electric power than the four ATM solar arrays could generate. *Skylab* could fulfill its scientific mission only if the first crew freed the remaining crippled solar wing. Using equipment that resembled long-handled pruning shears and a prybar, the astronauts pulled the stuck solar wing free. The space station was now ready to meet its scientific mission objectives. The duration of the first crewed mission was 28 days.

The second astronaut crew—Alan Bean, Jack Lousma, and Owen Garriott—was launched on July 28, 1973; mission duration on the space station was approximately 59 days and 11 hours. The third Skylab crew—Gerald Carr, William Pogue, and Edward Gibson—was launched November 16, 1973; mission duration was a little more than 84 days. Saturn IB rockets launched all three crews in modified Apollo spacecraft, which also served as their return-to-Earth vehicle. The third and final manned Skylab mission ended with splashdown in the Pacific Ocean on February 8, 1974.

The Skylab solar-experiment program provided much valuable data previously unavailable to solar scientists. Human crews operated *Skylab*'s Apollo Telescope Mount and made many useful scientific observations of the Sun. Taken collectively, these observations of the nearest star mark the beginning of a new era in space-based solar physics. Achieving finer detail in geometry and spectral resolution than had ever been seen before, the ATM probed the solar atmosphere from the far reaches of the corona to the lowest observable levels in the Sun's photosphere. Part of the long-term study of Skylab data included analysis of solar activity as it affects Earth's weather and communications.

After the last astronaut crew departed the space station in February 1974, it orbited Earth as an abandoned derelict. Unable to maintain its original altitude, the station finally reentered the atmosphere on July 11, 1979, during orbit 34,981. While most of the station was burned up during reentry, some pieces survived and impacted in remote areas of the Indian Ocean and sparsely inhabited portions of Australia.

✧ *Yohkoh* Spacecraft

The Japanese scientific spacecraft *Yohkoh* (meaning "sunbeam") was an 860-pound- (390-kg-) mass, solar X-ray observation satellite launched by the Japanese Institute of Space and Astronautical Sciences on August 30, 1991. The main objective of this satellite was to study the high-energy radiations from solar flares (i.e., hard and soft X-rays and energetic neutrons), as well as quiet Sun structures and pre–solar flare conditions.

Yohkoh was a three-axis stabilized observatory-type satellite that operated in a nearly circular orbit around Earth, carrying four instruments:

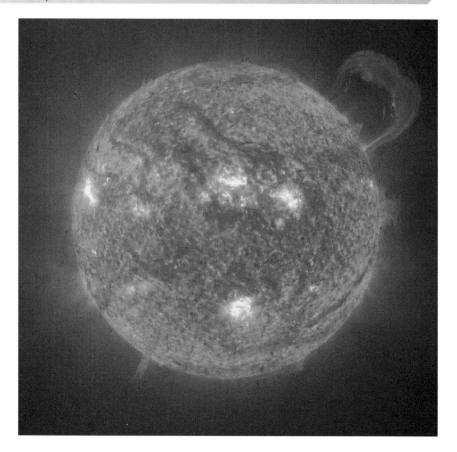

An extreme-ultraviolet imaging telescope image of a huge, handle-shaped prominence on the Sun taken by the *SOHO* spacecraft on September 14, 1999, at a wavelength of 304 angstroms (30.4 nm). Prominences are huge clouds of relatively cool, dense plasma suspended in the Sun's hot, thin corona. Emission in this spectral line shows the upper chromosphere is at a temperature of about 60,000 K. Every feature in the image traces magnetic field structure. The hottest appears almost white, while the darker areas indicate cooler temperatures. *(NASA, ESA, and SOHO/Extreme Ultraviolet Imaging Telescope [EIT] Consortium)*

two imagers and two spectrometers. The imaging instruments—a hard X-ray telescope (20–80 keV energy range) and a soft X-ray telescope (0.1–4 keV energy range)—had almost full-Sun fields of view to avoid missing any flares on the visible disk of the Sun.

This mission was a cooperative mission of Japan, the United States, and the United Kingdom. For example, *Yohkoh's* soft X-ray telescope was developed for NASA by the Lockheed Palo Alto Research Laboratory, in partnership with the National Astronomical Observatory of Japan and the Institute for Astronomy of the University of Tokyo. By the end of its mission in December 2001, the *Yohkoh* spacecraft had completed a decade of observing solar X-rays over the entire sunspot cycle. One unexpected result of the *Yohkoh* investigation was to show that the Sun's corona is much more active than scientists previously thought. In addition, the corona within active regions (that is, the sites of solar flares) was found to be expanding, in some cases almost continuously. Such expanding active regions apparently contribute to mass loss from the Sun and other stars.

✦ *Solar and Heliospheric Observatory*

The primary scientific aims of the European Space Agency's *Solar and Heliospheric Observatory* are to investigate the physical processes that form and heat the Sun's corona, maintain it, and give rise to the expanding solar wind; and to study the interior structure of the Sun. *SOHO* is part of the International Solar-Terrestrial Physics Program (ISTP) and involves NASA participation. The 2,970-pound (1,350-kg) (on-orbit dry mass) spacecraft was launched on December 2, 1995, and placed in a halo orbit at the Earth-Sun Lagrangian libration point one (L1) to obtain uninterrupted sunlight. It had a two-year design life, but onboard consumables were sufficient for an extra four years of operation. The spacecraft carried a complement of 12 scientific instruments.

In April 1998, *SOHO* successfully completed its nominal two-year mission to study the Sun's atmosphere, surface, and interior. The major science highlights of this mission included the detection of rivers of plasma beneath the observable surface of the Sun and the initial detection of solar flare–induced solar quakes. In addition, the *SOHO* spacecraft discovered more than 50 Sun-grazing comets. Then, on June 24, 1998, during routine housekeeping and maintenance operations, contact was lost with the *SOHO* spacecraft. After several anxious weeks, ground controllers were able to reestablish contact on August 3. They then proceeded to recommission various defunct subsystems and to perform an orbit correction maneuver. Their efforts brought *SOHO* back to a normal operational mode, with all its scientific instruments properly functioning by November 4, 1998.

✦ Ulysses Mission

The Ulysses mission is an international space robot designed to study the poles of the Sun and the interstellar environment above and below these solar poles. The *Ulysses* spacecraft is named for the legendary Greek hero in Homer's epic saga of the Trojan War who wandered into many

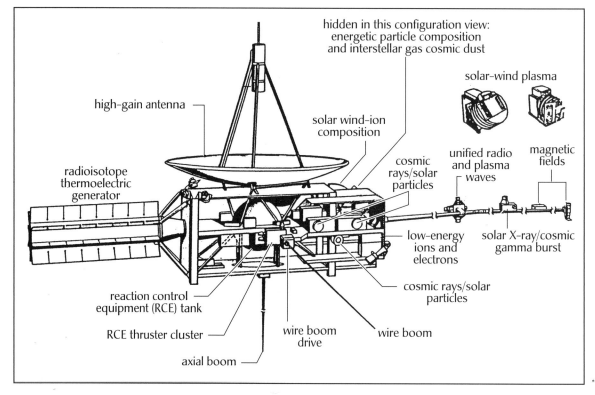

The compact *Ulysses* spacecraft and its array of scientific instruments *(Based on a drawing courtesy of NASA)*

unexplored areas on his return home. The spacecraft is on a survey mission designed to examine the properties of the solar wind, the structure of the Sun–solar wind interface, the heliospheric magnetic field, solar radio bursts and plasma waves, solar and galactic cosmic rays, and the interplanetary/interstellar neutral gas and dust environment—all as a function of solar latitude. Dornier Systems of Germany built the *Ulysses* spacecraft for the European Space Agency (ESA), which is responsible for in-space operations of the scientific mission.

NASA provided launch support using the space shuttle *Discovery* and an upper-stage configuration. In addition, the United States, through the Department of Energy, provided the radioisotope thermoelectric generator that supplies electric power to this spacecraft. *Ulysses* is tracked and its scientific data collected by NASA's Deep Space Network. Spacecraft monitoring and control, as well as data reduction and analysis, is performed at NASA's Jet Propulsion Laboratory (JPL) by a joint ESA/JPL team.

Ulysses is the first spacecraft to travel out of the ecliptic plane in order to study the unexplored region of space above the Sun's poles. To reach the necessary high solar latitudes, *Ulysses* was initially aimed close to Jupiter so that the giant planet's large gravitational field would accelerate the spacecraft out of the ecliptic plane to high latitudes. The gravitational assist encounter with Jupiter occurred on February 8, 1992. After the Jupiter encounter, *Ulysses* traveled to higher latitudes with maximum southern latitude of 80.2 degrees being achieved on September 13, 1994 (South Polar Pass 1).

Since *Ulysses* was the first spacecraft to explore the third dimension of space over the poles of the Sun, space scientists experienced some surprising discoveries. For example, they learned that two clearly separate and distinct solar-wind regimes exist, with fast wind emerging from the solar poles. Scientists were also surprised to observe how cosmic rays make their way into the solar system from galaxies beyond the Milky Way Galaxy. The magnetic field of the Sun over its poles turns out to be very different from previously expected—based on observations from Earth. Finally, *Ulysses* detected a beam of particles from interstellar space that was penetrating the solar system at a velocity of about 49,720 miles (80,000 km) per hour, or about 13.80 miles (22.22 km) per second.

Ulysses then traveled through high northern latitudes during June through September 1995 (North Polar Pass 1). The spacecraft's high-latitude observations of the Sun occurred during the minimum portion of the 11-year solar cycle.

In order to fully understand the Sun, however, scientists also wanted to study the parent star at near–maximum activity conditions of the 11-year cycle. The extended mission of the far-traveling, nuclear-powered scientific spacecraft provided the opportunity. During solar maximum conditions, *Ulysses* achieved high southern latitudes between September 2000 and January 2001 (South Polar Pass 2) and then traveled through high northern latitudes between September 2001 and December 2001 (North Polar Pass 2).

Now well into its extended mission, *Ulysses* continues to send back valuable scientific information on the inner workings of humans' parent star, especially concerning its magnetic field and how that magnetic field influences the solar system.

This mission was originally called the International Solar Polar Mission and was planned for two spacecraft, one built by NASA and the other by the ESA. However, NASA canceled its spacecraft part of the original mission in 1981 and instead provided launch and tracking support for the single spacecraft built by the ESA.

✧ Star Probe Mission

Star Probe is a conceptual robot spacecraft that can survive an approach to within about 1 million miles (1.6 million km) of the Sun's surface

This artist's concept shows a solar probe traveling within a million miles of the Sun. The robot spacecraft would use this visit to the nearest star to perform firsthand investigations of the physical conditions in the solar corona. The primary science objective of the solar probe is to help physicists understand the processes that heat the solar corona and produce the solar wind. *(NASA)*

(photosphere). This close encounter with the nearest star will give scientists their first direct (in situ) measurements of the physical conditions in the corona (the Sun's outer atmosphere). The challenging mission requires advanced robot spacecraft technologies, including superior thermal protection, specialized instrumentation, guidance and control, communications, and propulsion. NASA advanced mission planners suggest this type of robot spacecraft mission might be flown sometime between the years 2020 and 2030, with 2030 being the more conservative projection.

The science objectives of the mission include a determination of where and what physical processes heat the Sun's corona and accelerate the solar wind to its supersonic velocity. As now envisioned, the robot probe will combine remote sensing of the corona with in situ sampling within corona to produce a unique set of measurements not collected by any other spacecraft. Because of the extreme thermal environment in which the probe will have to operate, radioisotope thermoelectric generator units will be used as a reliable source of electric power throughout the mission. Gravity assist from the planet Jupiter will be used to give the three axes–stabilized spacecraft the final velocity it needs to fly on a trajectory very close to the Sun and sample the corona.

Star Probe will operate in hostile space environments over interplanetary distances ranging from about 0.2 astronomical unit (AU) to 5 AUs from the Sun. One astronomical unit corresponds to a distance of 93 million miles (149.6 million km). When the probe is at a distance of about 0.02 AU, the robot spacecraft will be just 1 million miles (1.6 million km) from the visible surface of the Sun (the photosphere) and approximately 1.4 million miles (2.3 million km) from the center of its thermonuclear-reacting core. When the probe is at five AU from the Sun, it will be traveling in close proximity to the planet Jupiter.

The Moon as a Platform for Astronomy and Astrophysics

When human beings return to the Moon this century, it will not be for a brief moment of scientific inquiry as occurred in NASA's Apollo Project, but rather as permanent inhabitants of a new world. They will build bases from which to completely explore the lunar surface, establish science and technology laboratories that take advantage of the special properties of the lunar environment, and harvest the Moon's resources (including the suspected deposits of lunar ice in the polar regions) in support of humanity's extraterrestrial expansion.

One of the most exciting concepts inherent in the establishment of a permanent lunar base is the ability of future scientists to efficiently and effectively use the Moon as a platform for advanced efforts in space-based astronomy and astrophysics. Of course robot platforms can be sent to the surface of the Moon before, or even in lieu of, the establishment of a permanent, human-occupied lunar base. However, the business of science—especially astronomy and astrophysics—could become one of the driving elements in the overall economy of a fledgling lunar settlement. This chapter will first introduce a generic lunar-base development scenario that spans the 21st century and then describe some concepts for astronomy performed on the surface of the Moon.

✦ Lunar-Base Scenarios and Concepts

A lunar base is a permanently inhabited complex on the surface of the Moon. In the first permanent lunar-base camp, a team of 10, up to perhaps 100, lunar workers will set about the task of fully investigating the Moon. The word *permanent* here means that the facility will always be occupied by human beings, but individuals probably will serve tours of from one to three years before returning to Earth. Some workers at the base will enjoy

160

being on another world. Some will begin to experience isolation-related psychological problems, similar to the difficulties often experienced by members of various scientific teams whose members "winter-over" in Antarctic research stations. Still other workers will experience injuries or even have fatal accidents while working at or around the lunar base.

For the most part, however, the pioneering lunar-base inhabitants will take advantage of the Moon as a science-in-space platform and perform the fundamental engineering studies needed to confirm and define the specific roles the Moon will play in astronomy, astrophysics, cosmology, and other scientific fields. The creation of a thriving "science business" on the Moon is an essential economic condition for the development of any self-sustaining settlement by the end of the century. The confirmation of frozen volatile resources (that is, lunar water ice) in the perpetually frozen recesses of the Moon's polar regions would significantly change lunar-base logistics strategies and could promote the development of a very large lunar settlement, containing up to 10,000 or more inhabitants.

In addition to using the Moon as a permanent platform for astronomy and astrophysics, many other lunar-base applications have been proposed. Some of these suggestions include: (1) a major lunar scientific laboratory complex (serving all fields of science); (2) a lunar industrial complex to support space-based manufacturing; (3) an asteroid- and comet-defense observatory performing solar system and deep-space surveillance; (4) a fueling station for orbital transfer vehicles that travel throughout cislunar and interplanetary space; and (5) a training site and assembly point for the first human expedition to Mars.

Social and political scientists suggest that a permanent lunar base could also become the site of innovative political, social, and cultural developments—essentially rejuvenating humans' concept of who people

This artist's rendering shows a three-foot- (1-m-) diameter transit telescope mounted to a robot lander spacecraft on the surface of the Moon. The instrument depicted here is called the lunar ultraviolet telescope experiment (LUTE). The proposed telescope would take advantage of the Moon as a stable, atmosphere-free platform and would also use the Moon's own slow rotation to perform a detailed telescopic survey of the extreme ultraviolet sky. *(NASA/JSC; artwork by John Frassanito and Associates [1992])* are as intelligent beings and boldly demonstrating the ability to beneficially apply advanced technology in support of the positive aspects of human destiny. Another interesting suggestion for a permanent lunar base is its use as a field operations center for the rapid-response portion of a planetary defense system that protects Earth from threatening asteroids or comets.

THE MOON

The Moon is Earth's only natural satellite and closest celestial neighbor. While life on Earth is made possible by the Sun, it is also regulated by the periodic motions of the Moon. For example, the months of the year are measured by the regular motions of the Moon around Earth, and the tides rise and fall because of the gravitational tug-of-war between Earth and the Moon.

Throughout history the Moon has had a significant influence on human culture, art, and literature. Even in the space age, it has proved to be a major technical stimulus. It was just far enough away to represent a real technical challenge to reach it, yet it was close enough to allow humans to be successful on the first concentrated effort. Starting in 1959 with the U.S. *Pioneer 4* and the Russian *Luna 1* lunar flyby missions, a variety of American and Russian missions have been sent to and around the Moon. The most exciting of these missions were the Apollo Project's human expeditions to the Moon from 1968 to 1972.

In 1994 the *Clementine* spacecraft, flown by the U.S. Department of Defense as a demonstration of certain advanced space technologies, spent 70 days in lunar orbit mapping the Moon's surface. Subsequent analysis of the *Clementine* data offered tantalizing hints that water ice might be present in some of the permanently shadowed regions at the Moon's poles. NASA launched the Lunar Prospector mission in January 1998 to perform a detailed study of the Moon's surface composition and to hunt for signs of the suspected deposits of water ice. Data from the Lunar Prospector mission strongly suggested the presence of water ice in the Moon's polar regions, although the results require additional confirmation. Water on the Moon (as trapped surface ice in the permanently shadowed regions of the Moon's poles) would be an extremely valuable resource that would open up many exciting possibilities for future lunar-base development.

From evidence gathered by the early robotic lunar missions (such as *Ranger*, *Surveyor*, and the *Lunar Orbiter* spacecraft), and by the Apollo missions, lunar scientists have learned a great deal more about the Moon and have been able to construct a geologic history dating back to its infancy.

Because the Moon does not have any oceans or other free-flowing water and lacks a sensible atmosphere, appreciable erosion—or weathering—has not occurred there. Consequently, scientists regard the Moon as a "museum world." The primitive materials that lay on its surface for billions of years are still in an excellent state of preservation. Scientists believe that the Moon was formed over 4 billion years ago and then differentiated quite early, perhaps only 100 million years later. Tectonic activity ceased eons ago on the Moon. The lunar crust and mantle are quite thick, extending inward to more than 500 miles (800 km). However, the deep interior of the Moon is still unknown. It may contain a small iron core at its center, and there is

As lunar activities expand, the original lunar base could grow into an early settlement of about 1,000 more or less permanent residents. Then, as the lunar industrial complex develops further and lunar raw materials, food, and manufactured products start to support space commerce throughout cislunar space, the lunar settlement itself will expand to a

some evidence that the lunar interior may be hot and even partially molten. Moonquakes have been measured within the lithosphere and interior, most being the result of gravitational stresses. Since the Moon appears seismically stable, very large structures or widely separated instruments placed on the surface for astronomical purpose should not experience any appreciable movement due to unwanted, naturally caused, ground movement.

Chemically, the Earth and the Moon are quite similar, although compared to Earth the Moon is depleted in more easily vaporized materials. The lunar surface consists of highlands composed of alumina-rich rocks that formed from a globe-encircling molten sea and maria made up of volcanic melts that surfaced about 3.5 billion years ago. However, despite all scientists have learned in the past three decades about Earth's nearest celestial neighbor, lunar exploration really has only just started. Several puzzling mysteries remain, including the origin of the Moon itself.

Recently, a new lunar origin theory has been suggested: a cataclysmic birth of the Moon. Scientists supporting this theory suggest that near the end of Earth's accretion from the primordial solar nebula materials (i.e., after its core was formed, but while Earth was still in a molten state), a Mars–size celestial object (called an "impactor") hit Earth at an oblique angle. This ancient explosive collision sent vaporized-impactor material and molten-Earth material into Earth's orbit, and the Moon then formed from these materials.

The surface of the Moon has two major regions with distinctive geologic features and evolutionary histories. First are the relatively smooth, dark areas that Galileo Galilei originally called maria (because he thought they were seas or oceans). Second are the densely cratered, rugged highlands (uplands) that Galileo Galilei called terrae. The highlands occupy about 83 percent of the Moon's surface and generally have a higher elevation (as much as three miles [5 km] above the Moon's mean radius). In other places, the maria lie about three miles (5 km) below the mean radius and are concentrated on the nearside of the Moon—that is, on the side of the Moon always facing Earth.

The main external geologic process modifying the surface of the Moon is meteoroid impact. Craters range in size from very tiny pits only micrometers in diameter to gigantic basins hundreds of miles across. The surface of the Moon is strongly brecciated, or fragmented. This mantle of weakly coherent debris is called regolith. It consists of shocked fragments of rocks, minerals, and special pieces of glass formed by meteoroid impact. Regolith thickness is quite variable and depends on the age of the bedrock beneath and on the proximity of craters and their ejecta blankets. Generally, the maria are covered by 9.8 feet (3 m) to 52.5 feet (16 m) of regolith, while the older highlands have developed a lunar soil at least 32.8 feet (10 m) thick. Some aerospace planners have identified contamination by lunar dust as a possible problem that needs to be considered in the design of optical equipment placed on the Moon's surface to support astronomy.

population of around 10,000. At that point, the original settlement might spawn several new settlements—each taking advantage of some special location or resource deposit elsewhere on the Moon's surface.

In the next (22nd) century, this collection of permanent human settlements on the Moon could continue to grow, reaching a combined population of about 500,000 persons and attaining a social and economic critical mass that supports true self-sufficiency from Earth. This moment of self-sufficiency for the lunar civilization will also be a very historic moment in human history. From that time on, the human race will exist in two distinct and separate biological niches—people will be terran (of Earth) and nonterran (or extraterrestrial).

With the rise of a self-sufficient, autonomous lunar civilization, future generations will have a choice of worlds on which to live and prosper. Of course such a major social development will most likely produce its share of cultural backlash in both worlds. Citizens of the next century may start seeing personal ground vehicles with such bumper-sticker slogans as, "Protect terrestrial astronomy jobs—ban outsourcing to lunar telescope facilities."

The vast majority of lunar-base development studies include the use of the Moon as a platform from which to conduct science in space. Scientific facilities on the Moon will take advantage of its unique environment to support platforms for astronomical, solar, and space science (plasma) observations. The unique environmental characteristics of the lunar surface include low gravity (one-sixth that of the Earth), high vacuum—about 10^{-12} torr (a torr is a unit of pressure equal to $1/760$ of the pressure of Earth's atmosphere at sea level)—seismic stability, low temperatures (especially in permanently shadowed polar regions), and a low radio-noise environment on the Moon's farside. This last attribute is especially important for future radio-astronomy activities.

A lunar scientific base also provides life scientists with a unique opportunity to extensively study biological processes in reduced gravity ($\frac{1}{6}$-g) and in low magnetic fields. Genetic engineers can conduct their experiments in comfortable facilities that are nevertheless physically isolated from the Earth's biosphere. Exobiologists can experiment with new types of plants and microorganisms under a variety of simulated alien-world conditions. Genetically engineered lunar plants, grown in special greenhouse facilities, could become a major food source, while also supplementing the regeneration of a breathable atmosphere for the various lunar habitats.

The true impetus for large, permanent lunar settlements will most likely arise from the desire for economic gain—a time-honored stimulus that has driven much technical, social, and economic development on Earth. The ability to create useful products from native lunar materials will have a controlling influence on the overall rate of growth of the lunar civilization. Some early lunar products can now easily be identified. Lunar ice, especially when refined into pure water or dissociated into the important chemicals hydrogen (H_2) and oxygen (O_2), represents the Moon's most important resource. Other important early lunar products include: (1) oxygen (extracted from lunar soils) for use as a propellant by orbital-transfer vehicles traveling throughout cislunar space; (2) raw (i.e., bulk, minimally processed) lunar soil and rock materials for space radiation shielding; and (3) refined ceramic and metal products to support the construction of large structures and habitats in space.

The initial lunar base can be used to demonstrate industrial applications of native Moon resources and to operate small pilot factories that provide selected raw and finished products for use both on the Moon and in Earth orbit. Despite the actual distances involved, the cost of shipping a kilogram of "stuff" from the surface of the Moon to various locations in cislunar space may prove much cheaper than shipping the same "stuff" from the surface of Earth.

The Moon has large supplies of silicon, iron, aluminum, calcium, magnesium, titanium and oxygen. Lunar soil and rock can be melted to make glass—in the form of fibers, slabs, tubes and rods. Sintering (a process whereby a substance is formed into a coherent mass by heating, but without melting) can produce lunar bricks and ceramic products. Iron metal can be melted and cast or converted to specially shaped forms using powder metallurgy. These lunar products would find a ready market as shielding materials, in habitat construction, in the development of large space facilities, and in electric power generation and transmission systems. These shielding materials are especially useful in the construction of very large and very sensitive, high-energy astrophysics instruments designed to study energetic cosmic rays or gamma rays.

Lunar mining operations and factories can be expanded to meet growing demands for lunar products throughout cislunar space. With the rise of lunar agriculture (accomplished in special enclosed facilities), the Moon may even become the "extraterrestrial breadbasket"—providing the majority of all food products consumed by humanity's extraterrestrial citizens.

Numerous other tangible and intangible advantages of lunar settlements will accrue as a natural part of their creation and evolutionary development. For example, the high-technology discoveries originating in a complex of unique lunar laboratories could be channeled directly into appropriate scientific, economic, and technical sectors on Earth, as so-called frontier ideas, techniques, products, and so on. The permanent presence of people on another world (a world that looms large in the night sky) will continuously suggest an open-world philosophy and a sense of cosmic destiny to the vast majority of humans who remain behind on the home planet. The human generation that decides to venture into cislunar space and to create permanent lunar settlements will long be admired, not only for its great technical and intellectual achievements, but also for its innovative cultural accomplishments. With a constant, unfiltered view of the stars, lunar-base inhabitants will always have a deep and warm affinity for astronomy and astrophysics. Finally, it is not too remote to speculate that the descendants of the first lunar settlers will become first the interplanetary, then the interstellar, portion of the human race. The Moon can be viewed as humanity's stepping-stone to the universe.

✦ Lunar Farside Radio Astronomy and Other Potential Astronomical Facilities

Radio astronomy is the branch of astronomy that collects and evaluates radio signals from extraterrestrial radio sources. Radio astronomy is a relatively young branch of astronomy. It was started in the 1930s, when Karl Guthe Jansky (1905–50), an American radio engineer, detected the first extraterrestrial radio signals. Until Jansky's discovery, astronomers had used only the visible portion of the electromagnetic spectrum to view the universe.

The detailed observation of cosmic radio sources is difficult, however, because these sources shed so little energy on Earth. But starting in the mid-1940s with the pioneering work of the British astronomer Sir Alfred Charles Bernard Lovell (b. 1913), at the United Kingdom's Nuffield Radio Astronomy Laboratories at Jodrell Bank, the radio telescope has been used to discover some extraterrestrial radio sources so unusual that their very existence had not even been imagined or predicted by scientists.

One of the strangest of these cosmic radio sources is the pulsar—a collapsed giant star that has become a neutron star and now emits pulsating radio signals as it spins. When the first pulsar was detected in 1967, it created quite a stir in the scientific community. Because of the regularity of its signal, scientists thought they had just detected the first interstellar signals from an intelligent alien civilization.

Another interesting celestial object is the quasar, or quasi-stellar radio source. Discovered in 1964, quasars are now considered to be entire galaxies in which a very small part (perhaps only a few light-days across) releases enormous amounts of energy—equivalent to the total annihilation of millions of stars. Quasars are the most distant-known objects in the universe; some of them are receding from Earth at over 90 percent of the speed of light.

This artist's rendering depicts a very large radio telescope facility that has been constructed within an impact crater basin on the farside of the Moon (circa 2030). Use of the Moon's farside shields the giant radio telescope (which as shown here is much larger than the Arecibo Observatory) from the unwanted human–produced radio-frequency signals that now flood the terrestrial environment. *(NASA)*

The Arecibo Observatory, a 1,000-foot- (305-m-) diameter giant metal dish, is the largest radio/radar telescope on Earth. The facility is located in a large, bowl-shaped natural depression in the tropical jungle of Puerto Rico. The Arecibo Observatory is the main observing instrument of the National Astronomy and Ionosphere Center (NAIC), a national center for radio and radar astronomy and ionospheric physics operated by Cornell University under contract with the National Science Foundation. The observatory operates on a continuous basis, 24 hours a day every day, providing observing time and logistic support to visiting scientists. When the giant telescope operates as a radio-wave receiver, it can listen for signals from celestial objects at the farthest reaches of the universe. As a radar transmitter/receiver, it assists astronomers and planetary scientists by bouncing signals off the Moon, off nearby planets and their satellites, asteroids, and even off layers of Earth's ionosphere.

The Arecibo Observatory has made many contributions to astronomy and astrophysics. In 1965 the facility (operating as a radar transmitter/ receiver) determined that the rotation rate of the planet Mercury was 59 days rather than the previously estimated value of 88 days. In 1974 the facility (operating as a radio-wave receiver) supported the discovery of the first binary pulsar system. This discovery led to an important confirmation of Albert Einstein's theory of general relativity and earned the American physicists Russell A. Hulse (b. 1950) and Joseph Hooten Taylor Jr. (b. 1941) the 1993 Nobel Prize in physics. In the early 1990s, astronomers used the facility to discover extrasolar planets in orbit around the rapidly rotating pulsar B1257+12.

Using the Arecibo Observatory as an example, several lunar-base planners have suggested selecting appropriate natural impact craters on the Moon in which to construct even larger radio telescopes. Radio astronomy from the lunar surface offers the distinct advantages of a low radio-noise environment and a stable platform in a low-gravity environment. The farside of the Moon is permanently shielded from direct terrestrial radio emissions. As future radio telescope designs approach their ultimate (theoretical) performance limits, this uniquely quiet lunar environment

An artist's rendering of an advanced robotic optical telescope on the lunar surface that uses a self-guided "walking" mobility platform. In one lunar-based astronomy scenario, several such identical telescopes communicate with each other and arrange themselves at appropriate locations and distances across the Moon's surface in order to function as a very large interferometer. *(NASA/JSC; artist, Pat Rawlings)*

may be the only location in all cislunar space where sensitive, radio-wave-detection instruments can be used to full advantage, both in radio astronomy and in the search for extraterrestrial intelligence (SETI). In fact, radio astronomy, including extensive SETI efforts, should represent one of the main lunar industries by the end of this century. In a certain sense, SETI performed by lunar-based scientists will be extraterrestrials searching for other extraterrestrials.

The Moon also provides a solid, seismically stable, low-gravity, high-vacuum platform for conducting precise interferometric and astrometric observations. An interferometer is an instrument that achieves high angular resolution by combining signals from at least two widely separated telescopes (optical interferometer) or a widely separated antenna array (radio interferometer). Radio interferometers are one of the basic instruments of radio astronomy. In principle, the interferometer produces and measures interference fringes from two or more coherent wave trains from the same source. These instruments are used to measure wavelengths, to measure the angular width of sources, to determine the angular position of sources, and for many other scientific purposes.

The Very Large Array (VLA) is an example of the type of extended radio-astronomy facility that can also be constructed in the lunar surface. The VLA is a spatially extended radio-telescope facility at Socorro, New Mexico. This facility consists of 27 antennae, each 82 feet (25 m) in diameter, that are configured in a giant "Y" arrangement on railroad tracks over a 12.4-mile (20-km) distance. The VLA is operated by the National Radio Astronomy Observatory and sponsored by the National Science Foundation.

The VLA has four major antenna configurations: A array, with a maximum antenna separation of 22.4 miles (36 km); B array, with a maximum antenna separation of 6.2 miles (10 km); C array, with a maximum antenna separation of 2.2 miles (3.6 km); and D array, with a maximum antenna separation of 0.6 mile (1 km). The operating resolution of the VLA is determined by the size of the array. At its highest, the facility has a resolution of 0.04 arc seconds. This corresponds, for example, to an ability to "see" a 43-gigahertz (GHz) radio-frequency source the size of a golf ball at a distance of 93 miles (150 km). The facility collects the faint radio waves emitted by a variety of interesting celestial objects and produces radio images of these objects with as much clarity and resolution as the photographs from some of the world's largest optical telescopes.

The technique of focusing and combining the signals from a distributed array of smaller telescopes to simulate the resolution of a single, much-larger telescope is called aperture synthesis. In astronomy a radio telescope is used to measure the intensity of the weak, static-like cosmic-radio waves coming from some particular direction in the universe.

ORBITING QUARANTINE FACILITY

The Orbiting Quarantine Facility is a proposed Earth-orbiting laboratory in which soil and rock samples from other worlds—for example, Martian soil and rock specimens—could first be tested and examined for potentially harmful alien microorganisms, before the specimens are allowed to enter Earth's biosphere. A space-based quarantine facility, either in stable orbit around Earth or on the surface of the Moon (in association with a permanent lunar base) provides several distinct advantages: (1) It eliminates the possibility of a sample-return spacecraft crashing and accidentally releasing its potentially deadly cargo of alien microorganisms; (2) it guarantees that any alien organisms that might escape from the orbiting laboratory's confinement facilities cannot immediately enter Earth's biosphere; and (3) it ensures that all quarantine workers remain in total isolation during protocol testing of the alien soil and rock samples.

Three hypothetical results of such protocol testing are: (1) no replicating alien organisms are discovered; (2) replicating alien organisms are discovered, but they are also found not to be a threat to terrestrial life-forms; or (3) hazardous, replicating alien life-forms are discovered. If potentially harmful replicating alien organisms were discovered during these protocol tests, then orbiting quarantine-facility workers would either render the sample harmless (for example, through heat- and chemical-sterilization procedures); retain it under very carefully controlled conditions in the orbiting complex and perform more detailed studies on the alien life-forms; or properly dispose of the sample before the alien life-forms could enter Earth's biosphere and infect terrestrial life-forms.

In addition to protocol testing, a quarantine facility associated with a future lunar base could provide scientists the unique opportunity of performing detailed studies on alien samples—perhaps even simulating alien-world environments in a type of extraterrestrial zoo. Exobiologists at the lunar base and on Earth (through telescience) could examine the alien life-forms (most likely microscopic organisms) without the fear of having a containment breach that then endangers the terrestrial biosphere. For further protection, the Moon-based quarantine facility could be located a significant distance away from the main base, with only authorized human personnel (or teleoperated robots) allowed inside the actual alien-world chambers.

Sensitivity is defined as the radio telescope's ability to detect these very weak radio signals, while resolution is defined as the telescope's ability to locate the source of these signals. The sensitivity of a distributed array of telescopes, such as the VLA, is proportional to the sum of the collecting areas of all the individual elements, while the array's resolution is determined by the distance (baseline) over which the array elements can be spread. Each of the 82-foot- (25-m-) diameter dish antennae in the VLA was specially designed with aluminum panels formed into a parabolic surface accurate to 0.02 inch (0.5 mm)—a design condition that enables the antennae to focus radio signals as short as one-centimeter wavelength.

The VLA is used to produce radio images with as much detail as those made by an optical telescope. To accomplish this, the VLA's 27 dish-shaped antennae are arranged in a giant Y-pattern, with the southeast and southwest arms of the Y-pattern each 13 miles (21 km) long, and the north arm 11.8 miles (19 km) long. The resolution of this radio telescope array is varied by changing the separation and spacing of its 27 antenna elements. The VLA is generally found in one of four standard-array configurations. In the smallest antenna-dispersion configuration (D array—the low-resolution configuration), the 27 individual antennae are clustered together and form an equivalent radio antenna with a baseline of just 0.62 mile (1 km). In the largest antenna-dispersion configuration (A array—the high-resolution configuration), the individual antennae stretch out in a giant Y-pattern that produces a maximum baseline of 22.4 miles (36 km).

Astrometry is the branch of astronomy that involves very precise measurements of the motion and position of celestial objects. For example, the availability of ultra-high-resolution (microarcsecond) optical, infrared, and radio observatories will allow astronomers to carefully search for extrasolar planets encircling nearby stars. (The search for extrasolar planets is discussed in chapter 11.)

With respect to planetary science and exobiology, the lunar base would also be an ideal location for an extraterrestrial quarantine facility. Astrobiology (also called exobiology) is the scientific discipline involving the search for and study of living organisms found on celestial bodies beyond Earth. Samples of possible life-bearing materials from Mars, Europa, or other solar system bodies, which are suspected of containing microorganisms potentially harmful to the terrestrial biosphere, could be analyzed, tested, challenged, and (if necessary) fully contained on the Moon. Under these circumstances, there would be a thorough scientific investigation of the samples by human scientists without exposing the terrestrial biosphere to any undo risk.

Searching for Extrasolar Planets, Brown Dwarfs, and Dark Matter

This chapter introduces three of the most interesting scientific quests in modern astronomy and discusses how space-based observatories, such as the *Hubble Space Telescope,* the *Chandra X-ray Observatory,* and the *Spitzer Space Telescope* are helping scientists in their quest to learn more about these elusive objects.

The first topic is that of extrasolar planets, that is, planets around other star systems. The big challenge to scientists today is finding and identifying Earth-like planets around Sun-like stars. If such planets are found, the next question is very obvious. Do such planets have life?

The next two topics are somewhat linked. Brown dwarfs are infrared (that is, relatively cool) celestial objects that are bigger than planets such as Jupiter but much smaller than stars such as the Sun. Astronomers and astrophysicists call such objects failed stars, meaning the object came together by gravity a long time ago but did not have enough mass to create core conditions that sustained thermonuclear burning of hydrogen. Astronomers searching for the missing mass or dark matter of the universe currently wonder whether there is a sufficient population of hard-to-detect brown dwarfs to account for the missing mass.

Astronomers call material in the universe that cannot be observed directly because it emits very little or no electromagnetic radiation, but whose gravitational effects can be measured and quantified, missing mass or dark matter. Accounting for this so-called dark matter will help astronomers and cosmologists make more accurate projections about the ultimate fate of the universe. Cosmology is the branch of science that deals with the origin, present state and, future of the universe. Chapter 12 discusses how human perception of the universe has changed through the ages as astronomy and astrophysics improved scientific understanding about celestial objects and phenomena.

✧ Extrasolar Planets

An extrasolar planet is a planet that belongs to a star other than the Sun. There are two general methods that can be used to detect extrasolar planets: direct, which involves a search for telltale signs of a planet's infrared emissions, and indirect, which involves precise observation of any perturbed motion (for example, wobbling) of the parent star or any periodic variation in the intensity or spectral properties of the parent star's light.

Growing evidence of planets around other stars is helping astronomers validate the hypothesis that planet formation is a normal part of stellar evolution. Detailed physical evidence concerning extrasolar planets—especially if scientists can determine their frequency of occurrence as a function of the type of star—would greatly assist scientists in estimating the cosmic prevalence of life. If life originates on suitable (Earth-like) planets whenever it can (as many exobiologists currently hold), then knowing how abundant such suitable planets are in the Milky Way would allow scientists to make more credible guesses about where to search for extraterrestrial intelligence and what the chances are of finding intelligent life beyond humans' own solar system.

Scientists have detected (through spectral light variation of the parent star) Jupiter-size planets around Sun-like stars, such as 51 Pegasi, 70 Virginis, and 47 Ursae Majoris. Detailed computer analyses of spectrographic data have revealed that light from these stars appears redder and then bluer in an alternating periodic (sine wave) pattern. This periodic light pattern could indicate that the stars themselves are moving back and forth along the line of sight possibly due to a large (unseen) planetary object that is

EARTH-LIKE PLANET

An Earth-like planet is an extrasolar planet that is located in an ecosphere and has planetary environmental conditions that resemble the terrestrial biosphere—especially a suitable atmosphere, a temperature range that permits the retention of large quantities of liquid water on the planet's surface, and a sufficient quantity of energy striking the planet's surface from the parent star. These suitable environmental conditions could permit the chemical evolution and the development of carbon-based life as scientists presently understand it on Earth. The planet also should have a mass somewhat greater than 0.4 Earth masses (to permit the production and retention of a breathable atmosphere) but less than about 2.4 Earth masses (to avoid excessive surface-gravity conditions).

An artist's rendering of a large extrasolar planet, known as a hot Jupiter, orbiting around a distant, alien star *(NASA)*

slightly pulling the stars away from (redder spectral data) or toward (bluer spectral data) Earth. The extrasolar planet around 51 Pegasi is sometimes referred to as a hot Jupiter, since it is a large planet (about half of Jupiter's mass) that is located so close to its parent star that it completes an orbit in just a few days (approximately 4.23 days). The star, 70 Virginis, is a main sequence star in the constellation Virgo. Astronomers have observed a very large object (about six to seven times the mass of Jupiter) orbiting this star. The large extrasolar planet, called 70 Virginis b, travels around its parent star at an average distance of one-half an astronomical unit (AU). The star, 47 Ursae Majoris, is a spectral type G1V yellow star (similar to the Sun) at a distance of about 46 light-years away from Earth. One of the first stars detected to possess an extrasolar planet, 47 Ursae Majoris is currently known to have two planets, called 47 Ursae Majoris b and c, respectively. 47 Ursae Majoris b is a massive planet (about 2.5 times Jupiter's mass) that travels in an almost circular orbit around its parent star at an average distance of approximately two AU. The second planet (47 Ursae Majoris c) has about 75 percent of Jupiter's mass and travels in a nearly circular orbit at an average distance of 3.7 AU from its parent star.

An artist's rendering of a suspected extrasolar planet. In May 2004, a clearing was detected around the star CoKu Tau 4 by NASA's *Spitzer Space Telescope*. Astronomers believe that an orbiting massive body (like the planet depicted here) may have swept away the star's disk material, leaving a central hole. *(NASA/JPL)*

To find extrasolar planets and characterize their atmospheres, scientists will use new (or planned) space-based observatories such as the *Spitzer Space Telescope*, the *James Webb Space Telescope*, and the *Kepler* spacecraft. For example, in late 2003, NASA's *Spitzer Space Telescope* captured a dazzling image of a massive disk of dusty debris encircling a nearby star called Fomalhaut. Scientists consider such disks to be remnants of planetary construction. Planetary scientists believe Earth formed out of a similar disk.

The *Spitzer Space Telescope* is now helping scientists identify other stellar dust clouds that might mark the sites of developing planets. In 2004 the infrared telescope gathered data to indicate a possible planet spinning its way through a clearing in a nearby star's dusty, planet-forming disk. *Spitzer* detected this clearing around the star CoKu Tau 4. Astronomers believe than an orbiting massive body, such as a planet, may have swept away the star's disk material, leaving a central hole. The possible planet is theorized to be at least as massive as Jupiter and may have a similar appearance to what the giant planets in this (humans') solar system looked like billions of years ago. As shown in the artist's rendering, a graceful ring spins high above the planet's cloudy atmosphere. The ring is formed from countless small, orbiting particles of dust and ice—leftovers from the initial gravitational collapse that formed the possible giant.

If some day human beings are able to visit an extrasolar planet like this one, they will have a very different view of the universe. The sky, instead of being the familiar dark expanse lit by distant stars, would be dominated by the thick disk of dust that fills this young planetary system. The view looking toward the alien solar system's parent star (CoKu Tau 4) would be relatively clear, because the dust in the interior disk has already fallen into the accreting star. A bright band would appear to surround the central star, caused by the central star's light being scattered back by the dust in the disc. Looking away from CoKu Tau 4, the dusty disk would appear dark, blotting out light from all the stars in the sky except those that lie well above the plane of the disk.

The *James Webb Space Telescope* (*JWST*) will be a large, single telescope that is folded to fit inside its launch vehicle and cooled to low temperatures in deep space to enhance its sensitivity to faint, distant objects. Mission controllers will operate *JWST* in an orbit far from Earth, away from the thermal energy (heat) radiated to space by our home planet. Scheduled for launch in 2011, one of this observatory's main science goals is to determine how planetary systems form and interact. The *JWST* can observe evidence of the formation of planetary systems (some of which may be similar to our own solar system) by mapping the light from the clouds of dust grains orbiting stars.

The star, Beta Pictoris, has such a cloud, as was discovered by the *Infrared Astronomical Satellite* (IRAS). These clouds are bright near the host stars and may be divided into rings by the gravitational influence of large planets. Scientists speculate that this dust represents material forming into planets. Around older stars, such dust clouds may be the debris of material that failed to condense into planets. The *JWST* will have unprecedented sensitivity to observe faint dust clouds around nearby stars. The infrared wavelength range is also the best way to search for planets directly, because they are brighter relative to their central stars. For example, at visible wavelengths, Jupiter is about 100 million times fainter than the Sun, but in the infrared, it is only 10,000 times fainter. A planet like Jupiter would be difficult to directly observe with any telescope on Earth, but the *JWST* has a chance to do so—because it operates in space beyond the disturbing influences of the terrestrial atmosphere.

Scheduled for launch in October 2008, NASA's *Kepler* spacecraft will use a unique space-based telescope specifically designed to search for Earth-like planets around stars beyond the solar system. This mission will allow scientists to search the Milky Way Galaxy for Earth-size or even smaller planets. The extrasolar planets discovered thus far are giant planets, similar to Jupiter, and are probably composed mostly of hydrogen and helium. Exobiologists believe that such Jupiter-sized planets are unlikely to harbor life. (These massive planets may, however, have moons with an atmosphere and liquid water on the surface. In such cases, life could arise

TRANSIT (PLANETARY)

A planetary transit involves the passage of one celestial body in front of another, much-larger-diameter celestial body. In solar system astronomy, one very important example is the transit of Venus across the face of the Sun, as seen by observers on Earth. Because of orbital mechanics, observers on Earth can witness only planetary transits of Mercury and Venus. There are about 13 transits of Mercury every century (100 years), but transits of Venus are much more rare. In fact, only seven such events have occurred since the invention of astronomical telescope. These transits took place in 1631, 1639, 1761, 1769, 1874, 1882, and, most recently, on June 8, 2004. Anyone who missed observing the 2004 transit should mark his or her astronomical calendar, because the next transit of Venus takes place on June 6, 2012.

Astronomers use contacts to characterize the principal events occurring during a transit. During one of the rare transits of Venus, for example, the event begins with contact I, which is the instant the planet's disk is externally tangent to the Sun. The entire disk of Venus is first seen at contact II, when the planet is internally tangent to the Sun. During the next several hours, Venus gradually traverses the solar disk at a relative angular rate of approximately four arc minutes per hour. At contact III, the planet reaches the opposite limb and is once again internally tangent to the Sun. The transit ends at contact IV, when the planet limb is externally tangent to the Sun. Contacts I and II define the phase of the transit called ingress, while astronomers refer to contacts III and IV as the egress phase or simply the egress.

From celestial mechanics, transits of Venus are only possible in early December and early June, when Venus's nodes pass across the Sun. If Venus reaches inferior conjunction at this time, a transit occurs. As noted from the list of the historic transits of Venus, these transits show a clear pattern of recurrence and take place at intervals of 8, 121.5, 8 and 105.5 years. The next eight-year pair of Venus transits will occur over a century from now on December 11, 2117, and December 8, 2125.

It is probably best to make plans to observe the upcoming June 6, 2012, transit of Venus. The entire 2012 transit (that is, all four contacts) will be visible from northwestern North America, Hawaii, the western Pacific, northern Asia, Japan, Korea, eastern China, the Philippines, eastern Australia, and New Zealand. Unfortunately, no portion of the 2012 transit will be visible from Portugal or southern Spain, western Africa, and the southeastern two-thirds of South America. For the parts of the world not previously mentioned, the Sun will be setting or rising while the transit is in progress—so it will not be possible to observe the complete event.

in these planetary systems. The Kepler mission is especially important because none of the planet-detection methods used to date have had the capability of finding Earth-size planets—those that are 30 to 600 times less massive than Jupiter. Furthermore, none of the giant extrasolar planets discovered to date probably has has liquid water on its surface or even a solid surface.

The Kepler mission is different from previous ways of looking for planets because it will look for the transit signature of planets. A transit occurs each time a planet crosses the line of sight between the planet's parent star and the observer. When this happens, the planet blocks some of the light from its star, resulting in a periodic dimming. This periodic signature is used to detect the planet and to determine its size and its orbit. Three transits of a star, all with a consistent period, brightness change, and duration, provide a robust method of detection and planet confirmation. The measured orbit of the planet and the known properties of the parent star are used to determine if each planet discovered is in the continuously habitable zone, that is, at the distance from its parent star where liquid water could exist on the surface of the planet. The *Kepler* spacecraft will hunt for planets using a specialized three-foot- (1-m-) diameter telescope called a photometer. This instrument can measure the small changes in brightness caused by the transits. By monitoring 100,000 stars similar to the Sun for four years following the *Kepler* spacecraft's launch, scientists expect to find many hundreds of terrestrial-type planets.

NASA's Terrestrial Planet Finder (TPF) mission consists of a suite of two complementary space observatories: a visible-light coronagraph and a mid-infrared formation-flying interferometer. An interferometer consists of a collection of several (small) telescopes that function together to produce an image much sharper than would be possible with a single telescope. As currently planned, the TPF coronagraph should launch in 2014 and the TPF interferometer by 2020. The combination will detect and characterize Earth-like planets around as many as 150 stars up to 45 light-years away. The science goals of this ambitious planet-hunting mission include a survey of nearby stars in the search for terrestrial-size planets in the continuously habitable zone. Scientists will then perform spectroscopy on the most promising candidate extrasolar planets, looking for atmospheric signatures characteristic of habitability or even life itself.

The challenge of finding an Earth-size (i.e., terrestrial) planet orbiting even the closest stars can be compared to finding a tiny firefly next to a blazing searchlight when both are thousands of miles away. The infrared emissions of a parent star are a million times brighter than the infrared emissions of any companion planets that might orbit around it. Beyond the year 2010, data from the Terrestrial Planet Finder should allow astronomers to analyze the infrared emissions of extrasolar planets in star systems up to about 100 light-years away. They will use these data to search for signs of atmospheric gases, such as carbon dioxide, water vapor, and ozone. Together with the temperature and radius of any detected planets, these atmospheric gas data will enable scientists to determine which extrasolar planets are potentially habitable, or even whether they may be inhabited by rudimentary forms of life.

The quest for extrasolar planets is one of the most interesting areas within modern astronomy. But how can astronomers find planetary bodies around distant stars? In their search for extrasolar planets, scientists employ several important techniques. These techniques include pulsar timing, Doppler spectroscopy, astrometry, and transit photometry.

Astronomers accomplished the first widely accepted detection of extrasolar planets (namely, pulsar planets) in the early 1990s using the pulsar-timing technique. Earth-mass and even smaller planets orbiting a pulsar were detected by measuring the periodic variation in the pulse arrival time. However, the planets detected were orbiting a pulsar—a dead star, rather than a dwarf (main-sequence) star. What is encouraging, however, about the detection is that the planets were probably formed after the supernova that resulted in the pulsar—thereby demonstrating that planet formation is probably a common rather than rare astronomical phenomenon.

Astronomers also use Doppler spectroscopy to detect the periodic velocity shift of the stellar spectrum caused by an orbiting giant planet. Scientists sometimes refer to this method as the radial velocity method. Using ground-based astronomical observatories, spectroscopists can measure Doppler shifts greater than about three meters per second due to the reflex motion of the parent star. This measurement sensitivity corresponds to a minimum detectable planetary mass equivalent to approximately 30 Earth masses for a planet located at one astronomical. This method can be used for main-sequence stars of spectral types mid-F through M. Stars hotter and more massive than mid-F rotate faster, pulsate, are generally more active, and have less spectral structure, thus making measurement of their Doppler shifts much more difficult. As previously mentioned, using this technique, scientists have successfully detected several large (Jupiter-like) extrasolar planets.

Scientists also use astrometry to look for the periodic wobble that a planet induces in the position of its parent star. The minimum detectable planetary mass gets smaller in inverse proportion to that planet's distance from the star. For a space-based astrometric instrument, such as NASA's planned Space Interferometry Mission (SIM)—a facility that will measure an angle as small as two micro–arc seconds—a minimum planet of about 6.6 Earth masses could be detected as it travels in a one-year orbit around a one solar-mass star that is 32.61 light-years (10 parsecs [pc]) from the

Earth. The SIM would also be capable of detecting a 0.4-Jupiter-mass planet in a four-year orbit around a star out to a distance of 1,630 light-years (500 pc). From the ground, modern telescope facilities such as the Keck Telescope (in Hawaii) can measure angles as small as 20 micro–arc seconds, leading to a minimum detectable planetary mass in a one AU orbit of 66 Earth masses for a one solar-mass star at a distance of 32.6 light-years (10 pc). The limitations to this method are the distance to the star and variations in the position of the photometric center due to star spots. There are only 33 nonbinary solar-like (F, G, and K) main-sequence stars within 32.6 light-years (10 pc) of the Earth. The farthest planet from its star that can be detected by this technique is limited by the time needed to observe at least one orbital period.

Finally, astronomers use the transit photometry technique to measure the periodic dimming of a star as caused by a planet passing in front of it along the line of sight from the observer. Stellar variability on the timescale of a transit limits the detectable size to about half that of Earth for a one AU orbit about a one-solar-mass star; or, with four years of observing, transit photometry can detect Mars-size planets in Mercury-like orbits. Mercury-size planets can even be detected in the continuously habitable zone of K and M stars. Planets with orbital periods greater than two years are not readily detectable, since their chance of being properly aligned along the line of sight to the star becomes very small.

Giant planets in inner orbits are detectable independent of the orbit alignment, based on the periodic modulation of their reflected light. The transit depth can be combined with the mass found from Doppler data to determine the density of the planet, as scientists have done in the case of a star called HD209458b. Doppler spectroscopy and astrometry measurements can be used to search for any giant planets that might also be in the systems discovered using photometry. Since the orbital inclination must be close to 90 degrees to cause transits, there is very little uncertainty in the mass of any giant planet detected. Photometry represents the only practical method for finding Earth-size planets in the continuously habitable zone.

✧ Brown Dwarfs

Bluntly stated, a brown dwarf is a failed star. The brown dwarf is a celestial object that forms through the contraction and fragmentation of an interstellar cloud in much the same way that real stars form and enter the main sequence. But the big difference is that somehow, the brown dwarf never reaches the necessary critical mass. This critical size (about 0.08 solar mass, or some 80 times the mass of Jupiter) is needed to allow sufficient gravitational contraction to create the intense temperature conditions (millions of kelvins) required for the ignition of hydrogen thermonuclear reactions in its core.

In other words, the brown dwarf is a cosmic dud. Originally just an interesting hypothesis put forth by astronomers to help explain where some of the universe's missing mass, or dark matter, might be, several candidate brown dwarfs have now been detected. The search is now on to determine the true population of these elusive, hard-to-detect celestial bodies.

Astronomers more formally classify the brown dwarf as a degenerate substellar (almost a star) object. It contains starlike material composition (that is, mostly hydrogen with some helium) but has too small a mass (generally between about 1 percent to 8 percent of a solar mass). The lack of sufficient mass prevents the core from initiating the thermonuclear fusion of hydrogen. Without hydrogen burning, the brown dwarf has a very low luminosity and is very difficult to detect.

Today astronomers use advanced infrared (IR) imaging techniques to find these unusual degenerate stellar objects, or failed stars. Some astronomers believe brown dwarfs make up a significant fraction of the missing mass (or dark matter) of the universe. In 1995 the first brown-dwarf

The illustration shows the approximate size of brown dwarf TWA 5B (center) compared to Jupiter (right) and the Sun (left). Although brown dwarfs are similar in size to Jupiter, they are much more dense and produce their own (typically infrared) light, whereas Jupiter shines with reflected light from the Sun. (NASA/CXC/M. Weiss)

candidate object (called Gliese 229B) was detected as a tiny companion orbiting the small red-dwarf star Gliese 229A. Since then astronomers have found other candidate brown dwarfs, using space-based telescopes.

While many candidate brown dwarfs have now been discovered in binary star systems, it appears that brown dwarfs are not being found as companions to Sun-like stars (those with about one solar mass). Rather, the data collected thus far, suggest that lower-mass (less than 0.5 solar mass and below, say) main-sequence stars are more likely to have a brown-dwarf companion. Astronomers remain somewhat uncertain about where to draw the line between K and M spectral class stars with small brown-dwarf companions and those with massive, Jupiter-sized objects (planets) as companions. Some astronomers suggest 12 Jupiter masses as the starting point for brown dwarf candidacy. At 12 Jupiter masses (and below), a celestial object (call it planet or a brown dwarf) with starlike material composition is not expected to experience nuclear fusion in its core. Above 12 Jupiter masses (but below 80 Jupiter masses), the substellar object may experience a brief episode of deuterium fusion in its contracting core, but the object's core never gets hot enough to achieve thermonuclear burning of hydrogen. Once any such deuterium fusion episode ends (due to the depletion of deuterium, a natural but relatively inabundant isotope of hydrogen), the brown dwarf loses its "thermonuclear option" forever. What remains is a degenerate substellar object that has essentially frozen somewhere in its pre–main sequence contraction phase. Astronomers have a difficult time observing brown dwarfs because the small and low temperature objects are often lost in the radiant glare of their main-sequence stellar companions. For example, astronomers used ground-based infrared observations to initially identify a candidate brown dwarf around Gliese 229A. They then confirmed their suspicions using follow-up observations by the *Hubble Space Telescope*.

✧ Dark Matter

Astronomers call material in the universe that cannot be observed directly because it emits very little or no electromagnetic radiation, but whose gravitational effects can be measured and quantified, dark matter. Dark matter was originally called missing mass and, as suggested by its name, was only discovered through its gravitational effects.

While investigating the cluster of galaxies in the Constellation Coma Berenices (Berenice's Hair) in 1933, the Swiss astrophysicist Fritz Zwicky (1898–1974) noticed that the velocities of the individual galaxies in this cluster were so high that they should have escaped from each other's gravitational attraction long ago. He concluded that the amount of matter actually present in the cluster had to be much greater than what could be accounted for by the visibly observable galaxies.

From his observations, Zwicky estimated that the visible matter in the cluster was only about 10 percent of the mass actually needed to gravitationally bind the galaxies together. He then focused a great deal of his own scientific attention on the problem of the missing mass of the universe. Today astronomers refer to this mystery as the problem of dark matter. Zwicky and other scientists used the rotational speeds of individual galaxies within a cluster of galaxies (as obtained from their Doppler shifts) to provide observational evidence that most of the mass of the universe might be in the form of invisible material called dark matter.

The second direct observational evidence of the existence of dark matter came from careful radio astronomy–supported studies of the rotation rate of individual galaxies, including the Milky Way Galaxy. From their rotational behavior, astronomers discovered that most galaxies appear to be surrounded by a giant cloud (or galactic halo)—containing matter capable of exerting gravitational influence but not emitting observable radiation. These studies also indicated that the great majority of a galaxy's mass lays in this very large halo, which is perhaps 10 times the diameter of the visible galaxy. The Milky Way is a large spiral galaxy that contains about 100 billion stars. Observations indicate that humans' home galaxy is surrounded by a dark-matter halo that probably extends out to about 750,000 light-years. The mass of this dark matter halo appears to be about 10 times greater than the estimated mass of all the visible stars in the Milky Galaxy. However, the dark-matter halo's material composition still remains an astronomical mystery. Using the Milky Way Galaxy as a reference, astrophysicists believe that the material contained in galactic halos could represent about 90 percent of the total mass of the universe—if the current gravity-only big bang cosmological model is correct.

It should come as no surprise that there is considerable disagreement within the scientific community as to what this dark matter really is. Two general schools of astronomical thought have emerged—one advocating MACHOs (or baryonic matter) and one advocating WIMPs (or nonbaryonic matter).

The first group assumes dark matter consists of MACHOs, or *massive compact halo objects*—essentially ordinary matter that astronomers have simply not yet detected. This unobserved, but ordinary, matter is composed of heavy particles (baryons), such as neutrons and protons. The brown dwarf is one candidate MACHO that could significantly contribute to resolving the missing mass problem. The brown dwarf is a substellar (almost a star) celestial body that has the material composition of a star but that contains too little mass to permit its core to initiate thermonuclear fusion. In 1995 astronomers detected the first brown-dwarf candidate object—a tiny companion orbiting the small red-dwarf star Gliese 229. Also very difficult to detect are low-mass white dwarfs that represent another dark-matter candidate. Using sophisticated observational techniques that take advantage of gravitational lensing, astronomers have collected data suggesting low-mass white dwarfs may actually make up about half of the dark matter in the universe. (Chapter 12 discusses gravitational lensing.) Finally, black holes, especially relatively low-mass primordial black holes that formed soon after the big bang, represent another MACHO candidate.

The second group of scientists speculates that dark matter consists primarily of exotic particles that they collectively refer to as WIMPs (*w*eakly *i*nteracting but *m*assive *p*articles). These exotic particles represent a hypothetical form of matter called nonbaryonic matter—matter that does not contain baryons (protons or neutrons). For example, if physicists determine that the neutrino actually possesses a tiny, nonzero rest mass, then these ubiquitous, but weakly interacting, elementary particles might carry much of the missing mass in the universe.

Some scientists suggest that the true nature of dark matter may not require an "all-or-nothing" characterization. To these astrophysicists, it seems perfectly reasonable that dark matter might exist in several forms, including difficult to detect low-mass stellar and substellar objects (MACHOs) in the inner regions of a dark-matter galactic halo, as well as swarms of WIMPs farther out in the galactic halo.

As presently understood, dark matter may be ordinary (but difficult to detect) baryonic matter; it may be in nonbaryonic form (such as possibly massive neutrinos); or perhaps it is an unexpected combination of both baryonic and nonbaryonic matter. In any event, current studies of the observable universe can account for only 10 percent of the mass needed to support gravity-only big bang cosmological models. Discovering the true nature of dark matter remains a very important and intriguing challenge for cosmologists and astrophysicists in this century.

Wrinkles in the Cosmic Microwave Background

Cosmology is the study of the origin, evolution, and structure of the universe. Contemporary cosmology centers around the big bang hypothesis—a theory stating that about 14 billion years ago the universe began in a great explosion and has been expanding ever since. In the open (or steady-state) model, scientists postulate that the universe is infinite and will continue to expand forever. In the more widely accepted closed universe model, the total mass of the universe is assumed sufficiently large to eventually stop its expansion and then make it start contracting by gravitation, leading ultimately to a big crunch. In the flat universe model, the expansion gradually comes to a halt, but instead of collapsing, the universe achieves an equilibrium condition with expansion forces precisely balancing the forces of gravitational contraction. The postulated presence of dark energy and its influence on an accelerating universe have tossed more linear-expansion cosmological models into a quandary.

Until the late 1990s, big bang cosmologists comfortably assumed that only gravitation was influencing the dynamics and destiny of the universe after the big bang explosion. Hubble's law seemed to be correct, although there was some disagreement about the value of Hubble's constant. Then, in 1998, astronomical observations of very bright supernovas indicated that the rate of expansion of the universe is actually increasing. These observations are now causing a great deal of commotion within the scientific community. The reason for this commotion is quite simple. Within the framework of contemporary big bang cosmology, scientists previously postulated that the gravitational pull exerted by matter in the universe slows the expansion imparted by the primeval big bang explosion. If the rate of expansion of the universe is actually increasing, what is counteracting the far-reaching and unrelenting force of gravity? Some scientists suggest that the universe could contain an exotic form of energy or matter that is gravitationally repulsive in its overall effect.

A resurrected and refashioned form of Einstein's cosmological constant could play an important role in describing the destiny of the universe. Before discussing how scientists this century try to estimate the impact on cosmology of a gravitationally repulsive cosmological constant (based on quantum vacuum energy), this chapter examines briefly how cosmology evolved from prehistory to the contemporary big bang model.

✧ Early Cosmologies

From ancient times, most societies developed one or more accounts of how the world (they knew) was created. These early stories are called creation myths. For each of these societies, their culturally based creation myth(s) attempted to explain (in very nonscientific terms) how the world started and where it was going.

In the second century C.E., Ptolemy, assembling and synthesizing all of early Greek astronomy, published the first widely recognized cosmological model, often referred to as the Ptolemaic system. In this geocentric cosmology model, Ptolemy codified the early Greek belief that Earth was at the center of the universe and that the visible planets (Mercury, Venus, Mars, Jupiter, and Saturn) revolved around Earth embedded on crystal spheres. The "fixed" stars appeared immutable (essentially unchangeable)—save for their gradual motions through the sky with the seasons—so they were located on a sphere beyond Saturn's sphere. While seemingly silly in light of today's scientific knowledge, this model could and did conveniently account for the motion of the planets then visible to the naked eye. Without detailed scientific data to the contrary, Ptolemy's model of the universe survived for centuries. Arab astronomers embraced and enhanced Ptolemy's work.

As a result of the efforts by Arab astronomers to preserve Greek cosmology, the notion of Earth as the center of the universe flowed back into western Europe as its people emerged from the Dark Ages and began to experience the Renaissance. Aristotle's teachings, including the unchallenged acceptance of geocentric cosmology, were integrated—almost as dogma—into a western European educational tradition and culture that drew heavily from faith-based teachings. Before the development of the scientific method (which accompanied the Scientific Revolution of the 17th century), even the most brilliant medieval scholars, such as Saint Thomas Aquinas (ca. 1225–74), did not feel the need to explain natural phenomena through repeatable experiments and physical laws capable of predicting the outcome of certain reactions. In a world where people used fire (combustion) and crude gunpowder rockets, but did not understand the basic principles behind thermodynamics, why should anyone challenge a comfortable Earth-centered model of the universe?

To the naked eye, the heavens appeared immutable (unchanging), except for the motion of certain fixed and wandering lights—which, once observed and monitored, became very useful in the development of calendars for religious, cultural, and economic activities. In the majority of evolved agrarian societies, astronomically based calendars became the best friend of farmers, alerting them to the pending change of seasons and guiding them when to plant certain crops. Farming-based civilizations were well served by the Ptolemaic system, so there was no pressing economic or social need to cause intellectual mischief. For example, suppose the world did move through space at some reckless speed, what would keep people and things from falling off? A round, rotating world was difficult enough to accept—but one that also moved rapidly through the emptiness of space, while spinning on its axis, was a very unsettling concept to many people in 16th-century western Europe.

✦ The Copernican Revolution

In 1543 a Polish church official turned astronomer, Nicholas Copernicus, upset the cosmic applecart. His book *De Revolutionibus Orbium Coelestium* (On the revolution of celestial spheres), published as he lay on his deathbed, boldly suggested that Earth was *not* the center of the universe; rather, the planet moved around the Sun just like the other planets. Copernican cosmology proposed a Sun-centered (heliocentric) universe and overturned two millennia of Greek astronomy. Church officials became very uncomfortable with Copernicus's book, but nothing really happened scientifically with its important message until the marriage of the telescope and astronomy in 1610. Even the brilliant naked-eye astronomer Tycho Brahe died a staunch advocate of geocentric cosmology, albeit his own refashioned version of the Ptolemaic system.

At the beginning of the Scientific Revolution in the early 17th century, the development of the astronomical telescope, the detailed observations of the planet Jupiter and its major moons by Galileo Galilei, and the emergence of Johann Kepler's laws of planetary motion provided the initial observational evidence and mathematical tools necessary to validate Copernican cosmology and to dislodge Ptolemaic cosmology. Copernicus and his advocacy of heliocentric cosmology triggered the start of modern science. However, the transition from the geocentric cosmology of Aristotle to the heliocentric cosmology of Copernicus did not take place without significant turmoil—especially within the religious sectors of European societies. At the time, previously homogeneous European Christianity was being torn asunder by the Protestant Reformation. Any idea that appeared to challenge traditional religious teachings had the potential for causing further instability in a world already torn apart by brutal warfare over disagreements about religious doctrine. So, if a person became identified as a Copernican, it often led to a high-profile public prosecution.

The trial of a Copernican usually resulted in an almost automatic conviction for heresy, followed by imprisonment or even execution. Despite such strong political and theological opposition, the Copernican system survived and emerged to rule cosmology for the next two centuries. The close study of the clockwork motion of the planets around the Sun and the Moon around Earth helped inspire Sir Isaac Newton when he created his universal law of gravitation (ca. 1680). Newton's concept of gravity proved valid only for bodies at rest or moving very slowly compared to the speed of light. However, these inherent limitations in Newtonian mechanics did not become apparent until the early part of the 20th century and the introduction of general relativity.

During the 18th and 19th centuries, improved mathematics, better telescopes, and the hard work of many dedicated scientists pushed Newtonian mechanics and its orderly model of the universe to its limit. Of course, there were a few gnawing questions that remained beyond the grasp of classical physics and the power of the optical instruments that were available at the time. For example, how big was the universe? Up until the third decade of the 20th century, most astronomers and cosmologists treated the universe and the Milky Way Galaxy as one and the same. The annoying little patches of fizzy light, nebulae, were generally regarded as groups of stars within the Milky Way that were simply beyond the optical-resolution limits of available instruments. A few scientists, such as Immanuel Kant (1724–1804), had already suggested the answer to the mystery. As discovered in the early 20th century, these fuzzy patches of light were really other galaxies—Kant's so-called island universes. What about the size of the universe? Well, as Edwin Hubble and others scientists eventually demonstrated, the universe is an incredibly large, expanding phenomenon filled with millions of galaxies rushing away from each other in all observable directions.

✧ Big Bang Cosmology

Modern cosmology has its roots in two major theoretical developments that occurred at the beginning of the 20th century. The first is the general theory of relativity, which Albert Einstein proposed in 1915. In it Einstein postulated how space and time can actually be influenced by strong sources of gravity. The subtle, but measurable, bending (warping) of a star's light as it passed behind the Sun during a 1919 solar eclipse confirmed that the gravitational force of a very massive object could indeed warp the space-time continuum.

After Einstein introduced his theory of general relativity, he, as well as a number of other scientists, tried to apply the new gravitational dynamics (that is, the warping of the space-time continuum) to the universe as a whole. At the time, this required the scientists to make an important theoretical assumption about how the matter in the universe was distributed. The simplest hypothesis they could make was to assume that, when viewed in any direction by different observers in any place, the universe would appear roughly the same to all of them. That is, the matter in the universe is assumed homogeneous and isotropic when averaged over very large scales. This important assumption is now called the cosmological principle, and it represents the second theoretical pillar of big bang cosmology. Space-based observatories, such as NASA's *Hubble Space Telescope,* now let astrophysicists and astronomers observe the distribution of galaxies on increasingly larger scales. The results continue to support the cosmological principle.

Furthermore, the cosmic microwave background—the remnant heat from the big bang explosion—has a temperature (about 2.7 K) that is highly uniform over the entire sky. This fact strongly supports the assumption that the intensely hot primeval gas, which emitted this radiation long ago, was very uniformly distributed.

Consequently, general relativity (linking gravity to the curvature of space-time) and the cosmological principle (the large-scale uniformity and homogeneity of the universe) form the entire theoretical basis for big bang cosmology and lead to specific predictions for observable properties of the universe.

The American astronomer Edwin Hubble provided the first important observation supporting big bang cosmology. During the 1920s, he performed precise observations of diffuse nebulae and then proposed that

COSMOLOGICAL PRINCIPLE

The cosmological principle is a hypothesis in modern cosmology that states that the expanding universe is isotropic and homogeneous. In other words, there is no special location for observing the universe, and all observers anywhere in the universe would see the same recession of distant galaxies. The cosmological principle implies the curvature of space–that is, since there is no center of the universe, there is no outer limit or surface.

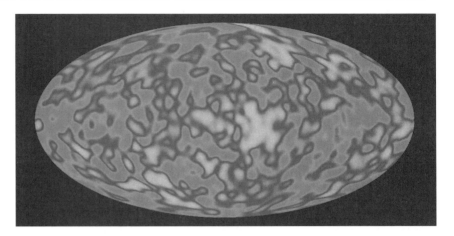

A microwave image of the entire sky, based on the first two years of differential microwave radiometer (DMR) data from NASA's *Cosmic Background Explorer* satellite. After computer processing to remove contributions from nearby objects and the effects of Earth's motion, this sky map shows the cosmic microwave background (CMB) anisotropy—tiny fluctuations in the sky brightness at the level of one part in 100,000. The CMB radiation is a remnant of the big bang, and the fluctuations are the imprint of density contrast in the early universe. *(NASA/GSFC)*

these diffuse objects were actually independent galaxies that were moving away from us in a giant, expanding universe. Modern cosmology, based on continuously improving astrophysical observations (observational and physical cosmology) and sophisticated theoretical developments (theoretical cosmology), was born.

In the 1940s, the Russian-American physicist George Gamow (1904–68) and others proposed a method by which Georges-Henri Lemaître's "cosmic egg" (i.e., the name for the initial big bang) could lead to the creation of the elements through nuclear synthesis and transformation processes in their presently observable cosmic abundances. This daring new cosmological model involved a giant explosion of an incredibly dense point object at the moment of creation. The so-called big bang was followed by a rapid expansion process, during which matter eventually emerged. The abundance of the light elements helium and hydrogen support big bang cosmology, because these light elements should have been formed from protons and neutrons when the universe was a few minutes old following the big bang.

Initially, the term *big bang* was sarcastically applied to this new cosmological model by rival scientists who favored a steady-state theory of the universe. In the steady-state cosmological model, the universe is assumed to have neither a beginning nor an end, and matter is thought to be added (created) continuously to accommodate the observed expansion of the galaxies. Despite the derisive origin of the name, both the name and the cosmological model that it represents have survived and achieved general acceptability.

Astrophysical discoveries throughout the 20th century tended to support the big bang cosmology—a model stating that about 14 billion years ago, the universe began in a great explosion (sometimes called the initial singularity) and has been expanding ever since. Physicists define a singularity as a point of zero radius and infinite density.

The 1964 discovery of the cosmic microwave background radiation by Arno Allen Penzias (b. 1933) and Robert Woodrow Wilson (b. 1936) provided the initial observational evidence that there was indeed a very-hot early phase in the history of the universe. More recently, space-based

WILKINSON MICROWAVE ANISOTROPY PROBE

NASA's *Wilkinson Microwave Anisotropy Probe (WMAP)* is designed to determine the geometry, content, and evolution of the universe through a high-resolution full-sky map of the temperature anisotropy of the cosmic microwave background (CMB). The choice of the spacecraft's orbit, sky-scanning strategy, and instrument design were driven by the science goals of the mission. The CMB sky-map data products derived from *WMAP*'s observations have 45 times the sensitivity and 33 times the angular resolution of NASA's Cosmic Background Explorer (COBE) mission.

In 1992 NASA's *COBE* satellite detected tiny fluctuations, or anisotropy, in the cosmic microwave background. This spacecraft found, for example, that one part of the sky has a CMB radiation temperature of 2.7251 K, while another part of the sky has a temperature of 2.7249 K. These fluctuations are related to fluctuations in the density of matter in the early universe and therefore carry information about the initial conditions for the formation of cosmic structures, such as galaxies, clusters, and voids.

The *COBE* spacecraft had an angular resolution of seven degrees across the sky–this is 14 times

(continues)

NASA's *Wilkinson Microwave Anisotropy Probe (NASA)*

(continued)

larger than the Moon's apparent size. Such limitations in angular resolution made *COBE* sensitive only to broad fluctuations in CMB of large size.

NASA launched the *Wilkinson Microwave Anisotropy Probe* in June 2001, and this spacecraft has been making sky maps of temperature fluctuations of the CMB radiation with much higher resolution, sensitivity, and accuracy than *COBE*. The new information contained in *WMAP*'s finer study of CMB fluctuations is shedding light on several very important questions in cosmology.

WMAP was launched by a Delta II rocket on June 30, 2001, from Cape Canaveral Air Force Station, Florida, and placed into a controlled Lissajous orbit about the second Sun-Earth Lagrange point (L2)—a distant orbit that is four times farther than the Moon and 0.93 million miles (1.5 million km) from Earth. Lissajous orbits are the natural motion of a satellite around a collinear libration point in a two-body system and require the expenditure of less momentum change for station keeping than halo orbits, where the satellite follows a simple circular or elliptical path around the libration point.

The *WMAP* spacecraft is expected to collect high-quality science data involving faint anisotropy or variations in the temperature of the CMB for at least four years. The *WMAP* spacecraft is mapping the cosmic microwave background at five frequencies: 23 gigahertz (GHz) (K-band), 33 GHz (Ka-band), 41 GHz (Q-band), 61 GHz (V-band), and 94 GHz (W-band). The CMB radiation that *WMAP* is observing is the oldest light in the universe and has been traveling across the universe for about 14 billion years.

observatories, such as NASA's *Cosmic Background Explorer* and the *Wilkinson Microwave Anisotropy Probe,* have provided detailed scientific data that not only generally support the big bang cosmological model but also raise interesting questions about it. For example, big bang cosmologists had to explain how the clumpy structures of galaxies could have evolved from a previously assumed "smooth" (that is, very uniform and homogeneous) big bang event.

The inflationary model of the big bang attempts to correlate modern cosmology and the quantum gravitational phenomena that are believed to have been at work during the very first fleeting moments of creation. This inflationary model suggests that the very early universe expanded so rapidly that the smooth homogeneity postulated in the original big bang model would be impossible. Although still being refined, the inflationary model appears to satisfy many of the perplexing inconsistencies that contemporary astrophysical observations had uncovered with respect to the more conventional big bang model. These refinements in big bang cosmology are expected to continue well into the 21st century, as even more sophisticated space observatories provide new data about the universe, its evolutionary processes, and its destiny.

✧ Cosmology in the 21st Century

In the late 1990s, two separate teams of astrophysicists examined very rare and bright stars called supernovas and then made the same startling announcement. Each team of scientists reported that the universe was expanding at an accelerating rate. Since then other scientists have gathered additional observational evidence to support the idea of cosmic acceleration. Pressed to explain what was going on and what could possibly be causing the universe to expand at a faster rate, cosmologist are now cautiously revisiting Einstein's cosmological constant and embracing a refashioned version of a gravitationally repulsive term—one that has its origins in quantum vacuum energy. Modern astronomical observations are also having profound implications for particle physics and could help scientists understand the fundamental forces of nature, especially the relation between gravity and quantum mechanics.

Today the main attraction of a cosmological constant term is that it significantly improves the agreement between theory and observation. The most interesting example of this is the recent effort to measure how much the expansion of the universe has changed in the last few billion years. Within the framework of contemporary big bang cosmology, scientists previously postulated that the gravitational pull exerted by matter in the universe slows the expansion. Now, if the universe is indeed expanding at an accelerating rate, they need to figure out what is causing this to occur. Cosmologists must now seriously consider that the universe may contain some bizarre form of matter or energy that is, in effect, gravitationally repulsive. The concept of a cosmological constant based on quantum vacuum energy is one candidate. But much work lies ahead before big bang cosmologists can be comfortable with suggesting the ultimate destiny of the expanding universe.

✧ The Fate of the Universe

One of the important roles of cosmology is to address the ultimate fate of the universe. In the open (or steady-state) model of the universe, scientists postulate that the universe is infinite and will continue to expand forever. In contrast, the closed universe model postulates that total mass of universe is sufficiently large, so that one day it will stop expanding and begin to contract due to the mutual gravitational attraction of the galaxies. This contraction will continue relentlessly until the total mass of the universe is essentially compressed into a singularity, known as the big crunch. Some advocates of the closed universe model also speculate that after the big crunch, there will be a new explosive expansion (that is, another big bang). This line of speculation leads to the pulsating or oscillating universe model—a cosmological model in which the universe appears and then disappears in an endless cycle between big bangs and big crunches. The concept of quintessence, a form of dark energy necessary to explain contemporary observations of an accelerating universe, further complicates the discussion by introducing the possibility of the big rip—an end state when the universe becomes gravitationally unbound.

Within big bang cosmology models whose dynamics are governed and controlled by gravity, the ultimate fate of the universe depends on the total amount of matter the universe contains. Does the universe contain enough matter to reverse its current expansion and cause closure? Astrophysical measurements of all observable luminous objects suggest that the universe contains only about 10 percent (or less) of the amount of mat-

DARK ENERGY

Dark energy is the current generic name given by astrophysicists and cosmologists to a hypothesized, unknown cosmic force field thought to be responsible for the recently observed acceleration of the expansion of the galaxies. In 1929 Edwin P. Hubble first proposed the concept of an expanding universe when he suggested that observations of Doppler-shifted wavelengths of the light from distant galaxies indicated that that these galaxies were receding from Earth with speeds proportional to their distance—an empirically based postulate that became known as Hubble's law. Mathematically, this relationship is expressed as: $V = H_0 \times D$, where V is the recessional velocity of a distant galaxy, H_0 is the Hubble constant, and D is its distance from Earth.

Then, in the late 1990s, while scientists made systematic surveys of very distant Type I (carbon detonation) supernovas, they observed that instead of slowing down (as might be anticipated if gravity

is the only significant force at work in cosmological dynamics), the rate of recession (that is, the redshift) of these very distant objects appeared to actually be increasing. It was almost as if some unknown force was neutralizing or canceling the attraction of gravity. Such startling observations proved controversial and very inconsistent with the standard gravity-only models of an expanding universe within big bang cosmology. However, despite fierce initial resistance within the scientific community, these perplexing observations eventually became accepted. The data imply that the expansion of the universe is accelerating—a dramatic conclusion that tosses modern cosmology into as great an amount of turmoil as did Hubble's initial announcement of an expanding universe some 70 years earlier.

But what could be causing this apparent acceleration of an expanding universe? Cosmologists do not yet have an acceptable answer. Some are revisiting the cosmological constant—a concept inserted ter thought to be needed to support the closed universe model. Where is the "missing mass" or dark matter? This is one important question that is perplexing modern scientists.

Cosmologists often discuss the mass of the universe in the context of the density parameter (symbol: Ω). It is a dimensionless parameter that expresses the ratio of the actual mean density of universe to the critical mass density—that is, the mean density of matter within the universe (considered as a whole) that cosmologists consider necessary, if gravitation is to eventually halt its expansion. One presently estimated value of this critical mass density is 8×10^{-27} kilograms per cubic meter, obtained by assuming that the Hubble constant (H_o) has a value of about 71.7 km/(s-Mpc [mega per second]) and that gravitation is all that is influencing the dynamic behavior of the expanding universe. If the universe does not contain sufficient mass (that is, if the $\Omega < 1$), then the universe will continue to expand forever. If $\Omega > 1$, then the universe has enough mass to eventually stop its expansion and start an inward collapse under the influence of gravitation.

by Albert Einstein into his general relativity theory to make revolutionary theory describe a static universe. However, after introducing the concept of a mysterious force associated with empty space capable of balancing or even resisting gravitational attraction, Einstein decided to abandon the concept. In fact, he referred to the notion of a cosmological constant as his "greatest failure." Nevertheless, contemporary physicists are now revisiting Einstein's concept and are suggesting that there is possibly a vacuum-pressure force (a recent name for the cosmological constant) that is inherently related to empty space but that exerts its influence only on a very large scale. Consequently, this mysterious force would have been negligible during the early phases of the universe following the big bang event, but it might now start to manifest itself and serve as a major factor in cosmological dynamics—that is, the rate of expansion of the present universe. Since such a mysterious force is neither required nor explained by any of the currently known laws of physics, scientists do not yet have a clear physical interpretation of just what such a mysterious (gravity resisting) force really means.

Today scientists grapple with various cosmological models in an effort to reconcile theory with challenging new data. To assist themselves in this process, physicists suggest the intriguing name "dark energy" to represent the mysterious cosmic force causing the universe to accelerate. Dark energy does not appear to be associated with either matter or radiation. What does the existence of dark energy mean? For one thing, the magnitude of the acceleration in cosmic expansion might imply that the amount of dark energy in the universe actually exceeds the total mass-energy equivalent of matter (luminous and dark) in the universe by a considerable margin. Any such (as yet unproven) dominant presence of dark energy—opposing the attractive force of gravity—then implies that the universe will continue to expand forever.

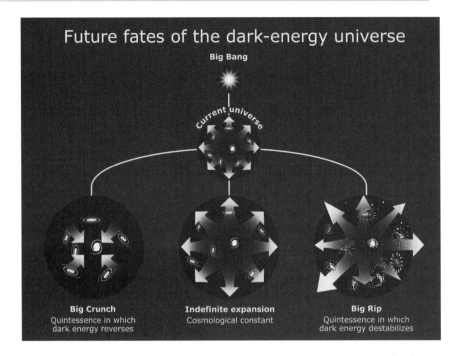

The illustration suggests the three possible fates of the universe, assuming big bang cosmology and the influence of dark energy: the big crunch, an ultimate collapse back into a singularity; indefinite expansion; and the big rip, in which the universe becomes gravitationally unbound and tears itself apart. Quintessence is a postulated form of dark energy considered necessary to explain observations of an accelerating universe. *(NASA and A. Field [STScI])*

If the critical mass density is just right (that is, if $\Omega = 1$), then the universe is considered flat, and a state of equilibrium eventually exists in which the outward force of expansion becomes precisely balanced by the inward force of gravitation. In the flat universe model, the expansion of the universe gradually comes to a halt but does not begin to collapse.

Today many cosmologists favor the inflationary theory embellishment of the big bang hypothesis (i.e., complex explosive birth) coupled with a flat universe ultimate fate. However, if an as-yet-undetected mysterious source of energy is now helping to overcome the attraction of gravity and causing the universe to expand at an accelerated rate, then its ultimate fate may be continued expansion—until all the galaxies and then all the stars fly away from each other. Cosmologists will have to work hard to unlock this mystery, the answer for which may lie in quantum gravitation and new interpretations of how the very rapid appearance and then disappearance of virtual particle-antiparticle pairs gives rise to quantum vacuum energy.

✦ Consciousness and the Universe

The history of the universe can also be viewed to follow a more or less linear timescale. This approach, sometimes called the scenario of cosmic evolution, links the development of the galaxies, stars, heavy elements, life, intelligence, technology, and the future. It is especially useful in philosophical and theological cosmology.

Exobiologists are also interested in understanding how life, especially intelligent life and consciousness, can emerge out of the primordial matter from which the galaxies, stars, and planets evolved. This cosmological approach leads to such interesting concepts as the living universe, the conscious universe, and the thinking universe. Is the evolution of intelligence and consciousness a normal endpoint for the development of matter everywhere in the universe? Or are human beings very unique by-products of the cosmic evolution process—perhaps the best the universe could do since the big bang? If the former question is correct, then the universe should be teeming with life, including intelligent life. If the latter is true, humans could be very much alone in a very big universe. If humans are the only beings now capable of contemplating the universe, then perhaps it is also our destiny to venture to the stars and carry life and consciousness to places where there is now only the material potential for such.

Up until recently, most scientists have avoided integrating the potential role of conscious intelligence (that is, matter that can think) into cosmological models. But what happens after conditions in the universe give rise to living matter that can think and reflect upon itself and its surroundings? The anthropic principle is an interesting, though highly controversial, hypothesis in contemporary cosmology, which suggests that the universe evolved after the big bang in just the right way so that life, especially intelligent life, could develop here on Earth. Does human intelligence (or possibly other forms of alien intelligence) have a role in the further evolution and destiny of the universe? If scientists and civilian authorities thought the Copernican hypothesis livened up intellectual activities in the 16th and 17th centuries, the revival of the cosmological constant and the bold hypothesis set forth in the anthropic principle should keep cosmologists very busy for a good portion of this century.

✦ Gravitation

Gravitation is a force in nature every person experiences, but it has also proven to be one of nature's most mysterious and interesting phenomena. In classical physics, Newton's universal law of gravitation defines gravitation as the force of mutual attraction experienced by two masses,

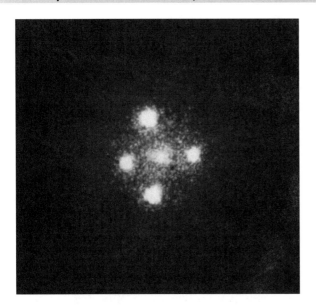

In 1990 the European Space Agency's Faint Object Camera onboard NASA's *Hubble Space Telescope* provided astronomers with a detailed image of the gravitational lens G2237 + 0305 (also called the "Einstein Cross"). This picture clearly depicts the interesting optical illusion caused by gravitational lensing. Shown are four images of the same distant quasar that has been multiple-imaged by a relatively nearby galaxy (central image) that acts as a gravitational lens. The angular separation between the upper and lower images is 1.6 arc seconds. *(NASA/ESA/STScI)*

directed along the line joining their centers of mass, and of magnitude inversely proportional to the square of the distance between the two centers of mass. According to Newton's theory of gravity, all masses pull on each other with an invisible force commonly called gravity. This force is an inherent property of matter and is directly proportional to an object's mass. In the solar system, the Sun reaches out across enormous distances and "pulls" smaller masses, such as planets, comets, and asteroids, into orbit around it, using its force of gravity (or gravitation).

In the early 20th century, Albert Einstein discovered a contradiction between Newton's theory of gravity and his (Einstein's) theory of special relativity, which he proposed in 1905. In special relativity, the speed of light is assumed to be the speed limit of all energy in the universe. No matter what kind of energy it is, it cannot propagate or transmit across the universe any faster than approximately 186,450 miles (300,000 km) per second. Yet, Newton's theory assumed that the Sun instantaneously transmits its force of gravity to the planets at a speed much faster than the speed of light. Einstein began to wonder whether gravity was unique in its ability to fly across the universe. He also explored another possibility, in thinking perhaps the masses reacted with each other for a different reason.

In 1916 Einstein published his general relativity theory—a theory that transformed space from the classic Newtonian concept of vast emptiness with nothing but the invisible force of gravity to rule the motion of matter, to an ephemeral fabric of space-time (a "space-time continuum") that enwraps and grips matter and directs its course through the universe. Einstein postulated that this space-time fabric spans the entire universe and is intimately connected to all matter and energy within it.

Theoretically, when a mass sits in the space-time fabric, it will deform the fabric itself, changing the shape of space and altering the passage of time around it. It is useful here to imagine the space-time fabric as a sturdy but flexible sheet of rubber stretched out to form a plane upon which an object like a baseball is placed. The ball causes the sheet to bend. Therefore, in Einstein's general relativity, a mass causes the fabric of space-time to

This artist's rendering is based on *Chandra X-ray Observatory* data from observations of RX J0806.3+1527 (or J0806), which show that the object's X-ray intensity varies with a period of 321.5 seconds. This implies that J0806 is a binary star system where two white dwarf stars are orbiting each other (as illustrated) approximately every five minutes. Because of the closeness of these two stars (about 50,000 miles [80,450 km] apart), they are moving in excess of a million miles (1.6 million km) per hour. According to Einstein's general relativity theory, this unusual binary star system should produce gravitational waves—ripples in space-time that carry energy away at the speed of light. *(NASA/GSFC/D. Berry)*

bend in a similar manner. The more massive the object, the deeper the dip in the fabric of space-time.

In the case of the Sun, for example, Einstein reasoned that because of its mass, the space-time fabric would curve around it, creating a "dip" in space-time. As the planets (as well as asteroids and comets) travel across the space-time fabric, they would respond to this Sun-caused dip and travel around the massive object by following the curvature in space-time. As long as they never slow down, the planets would maintain regular orbits around the Sun—neither spiraling in toward it nor flying off into interstellar space.

Einstein's concept of gravity has stimulated numerous scientific discussions, concepts, and challenges. For example, the graviton is the hypothetical elementary particle in quantum gravity that modern physicists speculate plays a role similar to that of the photon in quantum electrodynamics. Although experiments have not yet provided direct physical evidence of the existence of the graviton, Albert Einstein predicted the existence of a quantum (or particle) of gravitational energy as early as

1915, when he started formulating general relativity theory. Similarly, in modern physics, the gravitational wave is the gravitational analog of an electromagnetic wave whereby gravitational radiation is emitted at the speed of light from any mass that undergoes rapid acceleration.

Physicists are constantly probing Einstein's general relativity theory. One current aspect being studied is whether a heavy rotating object drags along the space-time continuum as its rotates. To answer that intriguing question, NASA constructed and orbited a very special spacecraft called *Gravity Probe B*. This NASA spacecraft was launched on April 20, 2004, by a Delta II rocket from Vandenberg Air Force Base, California. The scientific spacecraft, developed for NASA by Stanford University, is essentially a relativity gyroscope experiment designed to test two extraordinary, currently unverified predictions of Albert Einstein's general relativity theory.

Gravity Probe B's experiment is very precisely investigating tiny changes in the axis of spin of four gyroscopes contained in a satellite that orbits at an altitude of 398 miles (640 km) directly over Earth's poles. So free are these special gyroscopes from disturbance that they provide an almost perfect space-time reference system. Specifically, they are allowing scientists to measure how the fabric of space-time is being warped by the presence of Earth. Even more profoundly, *Gravity Probe B* is also allowing scientists to measure how Earth's rotation drags space-time around with it. Confirmation of these anticipated general relativity effects, though small for an object the mass of Earth, would have far-reaching implications for the nature of matter and the structure of the universe. NASA expects to release the scientific results of the Gravity Probe B mission in late 2006.

Aerospace technicians inspect NASA's *Gravity Probe B* spacecraft at Vandenberg Air Force Base in California. *Gravity Probe B* was successfully launched on April 20, 2004. *(NASA/MSFC)*

✧ Black Holes

One of the most fascinating objects in astronomy and astrophysics is the black hole, a gravitationally collapsed mass from which nothing—light, matter, or any other kind of signal—can escape.

Astrophysicists now conjecture that a stellar black hole is the natural end product when a giant star dies and collapses. If the core (or central region) of the dying star has three or more solar masses left after exhausting its nuclear fuel, then no known force can prevent the core from forming the extremely deep (essentially infinite) gravitational warp in space-time known as a black hole. As with the formation of a white dwarf or a neutron star, the collapsing giant star's density and gravity increase with gravitational contraction. However, in this case, because of the larger mass involved, the gravitational attraction of the collapsing star becomes too strong for even neutrons to resist, and an incredibly dense point mass, or singularity, forms. Physicists define this singularity as a point of zero radius and infinite density—in other words, it is a point where space-time is infinitely distorted. Basically, a black hole is a singularity surrounded by an event region in which the gravity is so strong that absolutely nothing can escape.

No one will ever know what goes on inside a black hole, since no light or any other radiation can escape. Nevertheless, physicists try to study the black hole by speculating about its theoretical properties and then looking for perturbations in the observable universe that provide telltale signs that a massive, invisible object matching such theoretical properties is possibly causing such perturbations. For example, here on Earth, a pattern of ripples on the surface of an otherwise quiet, but murky, pond could indicate that a large fish is swimming just below the surface. Similarly, astrophysicists look for detectable ripples in the observable portions of the universe to support theoretical predictions about the behavior of black holes.

The very concept of a mysterious black hole exerts a strong pull on both scientific and popular imaginations. Data from space-based observatories, such as the *Chandra X-ray Observatory,* have moved black holes from the purely theoretical realm to a dominant position in observational astrophysics. Strong evidence is accumulating that black holes not only exist but that very large ones called supermassive black holes, which contain millions or billion of solar masses, also may function like cosmic monsters lurking at the centers of every large galaxy.

How did the idea of a black hole originate? The first person to publish a paper about black holes was John Michell (1724–93), a British geologist, amateur astronomer, and clergyman. His 1784 paper suggested the possibility that light (then erroneously believed to consist of tiny particles of matter subject to influence by Newton's law of gravitation) might not be able to escape from a sufficiently massive star. Michell was a competent

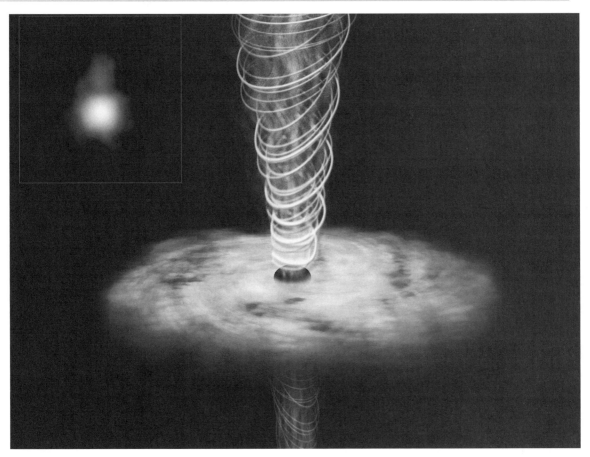

An artist's rendering of quasar GB1508+5714 and the most distant X-ray jet yet discovered. In 2003 a *Chandra X-ray Observatory* image of the quasar (inset on the left) revealed a jet of high-energy particles that extends more than 100,000 light-years from the supermassive black hole powering the quasar GB1508+5714. At a distance of 12 billion light-years from Earth, this is the most distant jet ever detected. *(NASA/CXC/A. Siemiginowska et al; illustration: CXC/M. Weiss)*

astronomer who successfully investigated binary star-system populations. He further suggested in this 18th-century paper that, although no "particles of light" could escape from such a massive object, astronomers might still infer its existence by observing the gravitational influence it exerted on nearby celestial objects. The French mathematician and astronomer Pierre-Simon, marquis de Laplace (1749–1827) introduced a similar concept in the late 1790s, when he also applied Newton's law of gravitation to a celestial body so massive that the force of gravity prevented any light particles from escaping. However, neither Laplace nor Michell used the term *black hole* to describe their postulated very-massive heavenly body. In fact, the term did not enter the lexicon of astrophysics until the American physicist John Archibald Wheeler (b. 1911) introduced it in 1967. Both of these 18th-century black-hole speculations were on the right track but suffered from incomplete and inadequate physics.

The needed breakthrough in physics took place a little more than a century later, when Albert Einstein introduced relativity theory. Einstein replaced the Newtonian concept of gravity as a force with the notion that gravity was associated with the distortion of the space-time. As originally suggested by Einstein, the more massive an objective, the greater its ability to distort the local space and time continuum.

Shortly after Einstein introduced general relativity in 1915, the German astronomer Karl Schwarzschild (1873–1916) discovered that Einstein's general relativity equations led to the postulated existence of a dense object into which other objects could fall, but out of which no objects could ever escape. In 1916 Schwarzschild wrote the fundamental equations that describe a black hole. He also calculated the size of the event horizon, or boundary of no return, for this incredibly dense and massive object. The dimension of the event horizon now bears the name Schwarzschild radius in his honor.

The notion of an event horizon implies that no information about events taking place inside this distance can ever reach outside observers. However, the event horizon is not a physical surface. Rather, it represents the start of a special region that is disconnected from normal space and time. Although scientists cannot see beyond the event horizon into a black hole, they believe that time stops there.

Inside the event horizon, the escape speed exceeds the speed of light. Outside the event, escape is possible. It is important to remember that the event horizon is not a material surface. It is the mathematical definition of the point of no return, the point where all communication is lost with the outside world. Inside the event horizon, the law of physics as humans know them does not apply. Once anything crosses this boundary, it will disappear into an infinitesimally small point—the singularity, as previously mentioned. Scientists cannot observe, measure, or test a singularity, since it, too, is a mathematical definition. The little that scientists currently know about black holes comes from looking at the effects they have on their surroundings *outside* the event horizon. As even more powerful space-based observatories study the universe across all portions of the electromagnetic spectrum in this century, scientists will be able to construct better and better theoretical models of the black hole.

The mass of the black hole determines the extent of its event horizon. The more massive the black hole, the greater the extent of its event horizon. The event horizon for a black hole containing one solar mass is just 1.9 miles (3 km) from the singularity. This is an example of a stellar black hole. In the Milky Way galaxy, a supernova occurs an average of once or twice every 100 years. Since the galaxy is about 1 billion years old, scientists suggest that about 200 million supernovas have occurred, creating neutron stars and black holes. Astronomers have identified nearly a dozen stellar black-hole candidates (containing between three and 20 or more solar masses) by observing how some stars appear to wobble as if nearby, yet invisible, massive companions were pulling on them. X-ray binary systems offer another way to search for candidate black holes.

In comparison to stellar black holes, the event horizon for a supermassive black hole consisting of 100 million solar masses is about 300 million kilometers from its singularity—twice the distance of the Sun to Earth. Astronomers suspect that such massive objects exist at the centers of many galaxies because they provide one of the few logical explanations for the strange and energetic events now being observed there. No one knows for sure how such supermassive black holes formed. One hypothesis is that over billions of years, relatively small stellar black holes (formed by supernovas) began devouring neighboring stars in the star-rich centers of large galaxies and eventually became supermassive black holes.

Once matter crosses the event horizon and falls into a black hole, only three physical properties appear to remain relevant: total mass, net electric charge, and total angular momentum. Recognizing all black holes must have mass, physicists have proposed four basic stellar black-hole models. The Schwarzschild black hole (first postulated in 1916) is a static black hole that has no charge and no angular momentum. The Reissner-Nordström black hole (introduced in 1918) has an electric charge but no angular momentum—that is, it is not spinning. In 1963 the New Zealand mathematician Roy Patrick Kerr (b. 1934) applied general relativity to describe the properties of a rapidly rotating, but uncharged, black hole. Astrophysicists think this model is the most likely "real-world" black hole because the massive stars that formed them would have been rotating. One postulated feature of a rotating Kerr black hole is its ringlike structure (the ring singularity) that might give rise to two separate event horizons. Some daring astrophysicists have even suggested it might become possible (at least in theory) to travel through the second event horizon and emerge into a new universe or possibly a different part of this universe. The final black-hole model has both charge and angular momentum. Called the Kerr-Newman black hole, this theoretical model appeared in 1965. However, astrophysicists currently think that rotating black holes are unlikely to have a significant electric charge, so the uncharged Kerr black hole remains the more favored "real-world" candidate model for the stellar black hole.

Current astrophysical evidence that superdense stars, such as white dwarfs and neutron stars, really exist supports the theoretical postulation that black holes themselves—representing the ultimate in density—must also exist. But how can scientists detect an object from which nothing, not even light, can escape? Astrophysicists think they may have found indirect ways of detecting black holes. The best currently available techniques depend upon candidate black holes being members of binary star systems. Unlike the Sun, many stars (over 50 percent) in the Milky Way Galaxy are members of a binary system. If one of the stars in a particular binary system has become a black hole, although invisible, it would betray its existence by the gravitational effects it produces upon the observable companion star. Once beyond event horizon, the black hole's gravitational influence is the same as that exerted by other objects (of equivalent mass) in the "normal" universe. So a black hole's gravitational effects on its companion would obey Newton's universal law of gravitation—that is, the mutual gravitational attraction of the two celestial objects is directly proportional to their masses and inversely proportional to the square of the distance between them. Astrophysicists have also speculated that a substantial part of the energy of matter spiraling into a black hole is converted by collision, compression, and heating into X-rays and gamma rays that display certain spectral characteristics. X-ray and gamma radiations emanate from the material as it is pulled toward the black hole. However, once the captured material crosses the black hole's event horizon, this telltale radiation cannot escape.

Suspected black holes in binary star systems exhibit this type of prominent material capture effect. Astronomers have discovered several black-hole candidates, using space-based astronomical observatories (such as the *Chandra X-ray Observatory*). One very promising candidate is called Cygnus X-1, an invisible object in the constellation Cygnus (The swan). The notation Cygnus X-1 means that it is the first X-ray source discovered in Cygnus. X-rays from the invisible object have characteristics like those expected from materials spiraling toward a black hole. This material is apparently being pulled from the black hole's binary companion, a large star of about 30 solar masses. Based upon the suspected black hole's gravitational effects on its visible stellar companion, the black hole's mass has been estimated to be about six solar masses. In time the giant visible companion might itself collapse into a neutron star or a black hole; or else it might be completely devoured by its black-hole companion. This form of stellar cannibalism would also significantly enlarge the existing black hole's event horizon.

In 1963 the Dutch-American astronomer Maarten Schmidt (b. 1929) was analyzing observations of a "star" named 3C 273. He had very confusing optical and radio data. What he and his colleagues had discovered was the quasar. Today astronomers know that the quasar is a type of active galactic nucleus (AGN) in the heart of a normal galaxy. AGN galaxies have hyperactive cores and are much brighter than normal galaxies. The AGNs emit energy equivalent to converting the entire mass of the Sun into pure energy every few years. Energy is emitted in all regions of the electromagnetic spectrum, from low-energy radio waves all the way to much-higher-energy X-rays and gamma rays. Furthermore, the energy output of these AGNs can vary on short timescales (hours and days), suggesting that the source is very compact. Stars by themselves, powered by nuclear fusion, cannot generate such levels of energy. Even the impressive supernova explosions are insufficient. So astrophysicists puzzle over what physical processes could produce the power of more than 100 Milky Way Galaxies and do it within a region of space only a few light-years across.

To further compound this cosmic mystery, some of these AGN galaxies have extraordinary jets of material rushing out of their cores that stretch far into space—up to 100 to 1,000 times the diameter of the galaxy. After considering all types of energetic processes, including the simultaneous explosions of thousands of supernovas, most astrophysicists now think that supermassive black holes represent the most plausible answer. Although nothing emerges from a black hole, matter falling into one can release tremendous quantities of energy just before it crosses the event horizon. For example, the region just outside the event horizon will glow in X-rays and gamma rays—the most energetic forms of electromagnetic radiation.

Matter captured by the gravity of a black hole will eventually settle into a disk around the black hole. Scientists call the inner region of swirling, superheated material an accretion disk. Stellar black holes can have accretion disks if they have a nearby companion star. Material from the companion will be drawn into orbit around the black hole, thereby forming an accretion disk. The diameter of the accretion disk depends on the mass of the black hole. The more massive the black hole, the larger the accretion disk. The accretion disk of a stellar black hole will stretch out only a few hundred or thousand kilometers from the center. However, the accretion disk of a supermassive black hole is much bigger and becomes solar system–sized.

Perhaps the most spectacular accretion disks exist in active galaxies that probably contain supermassive black holes. The *Hubble Space Telescope* has provided astronomers with strong evidence for this assumption. For example, the galaxy NGC 4261 is an elliptical galaxy whose core contains an unexpected large disk of dust and gas. Astronomers think that a black hole may lurk within the central region of this galaxy. Radio observations of this galactic core have also revealed jets of material ejected from the center of this disk. This provides corroborating evidence for the existence of a very large black hole.

Exactly what creates and controls the flow of matter from these jets is still not clearly understood. It is almost as if black holes are messy eaters, consuming matter, but spewing out leftovers. Most likely the jets have something to do with the rotation and/or the magnetic fields of the black hole. Whatever the actual cause, most astronomers believe that only black holes are capable of producing such spectacular and outrageous behavior.

Using the *Hubble Space Telescope,* scientists have also discovered a 3,700-light-year-in-diameter dust disk encircling a suspected 300-million solar-mass black hole in the center of the elliptical galaxy NGC 7052, located in the constellation Vulpecula about 191 million light-years from Earth. This disk is thought to be the remnant of an ancient galaxy collision that will be swallowed up by the giant black hole in several billion years. *Hubble Space Telescope* measurements have shown that the disk rotates rapidly at 96 miles (155 km) per second at a distance of 186 light-years from the center. The speed of rotation provides scientists a direct measure of the gravitational forces acting on the gas due to the presence of a suspected supermassive black hole. Though 300 million times the mass of the Sun, this suspected black hole is still only about 0.05 percent of the total mass of the NGC 7052 galaxy. The bright spot in the center of the giant dust disk is the combined light of stars that have been crowded around the black hole by its strong gravitational pull. This stellar concentration appears to match the theoretical astrophysical models that link stellar density to a central black hole's mass.

In the 1990s, X-ray data from the German-American *Roentgen Satellite* and the Japanese-American *Advanced Satellite for Cosmology and Astrophysics* suggested that a mid-mass black hole might exist in the galaxy M82. This observation was confirmed in September 2000, when astronomers compared high-resolution *Chandra X-ray Observatory* images with optical, radio, and infrared maps of the region. Scientists now think such black holes must be the results of black-hole mergers, since they are far too massive to have been formed from the death of a single star. Sometimes called the "missing-link" black holes, these medium-size black holes fill the gap in the observed (candidate) black-hole masses between stellar and supermassive. The M82 "missing link" is not in the absolute center of the galaxy where all supermassive black holes are suspected of residing, but it is comparative close to it.

The detailed, multispectral study of the environmental influence exerted by a suspected black hole on the nearby universe just beyond its event horizon is an exciting area of contemporary space-based astronomy. Unanticipated discoveries made in the next few decades could easily change the trajectory of contemporary physics—influencing, in turn, the trajectory of human civilization in the process.

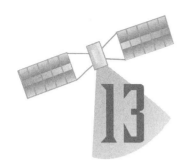

Conclusion

In 1610 Galileo Galilei's simple optical telescope caused a revolution in observational astronomy that promoted the companion revolution in scientific thinking that transformed Western civilization. Today the detailed, multispectral study of the universe made possible by a variety of powerful space-based astronomical observatories is dramatically changing the practice of contemporary science on a short timescale that is without precedence in human history. To many scientists and nonscientists alike, the only thing constant in their lives appears to be change.

When this book began to be written, for example, astronomers comfortably taught that the planet Pluto had one large moon, named Charon. As this manuscript approached completion, astronomers announced that they had used data from the *Hubble Space Telescope* to detect two additional smaller moons around that frigid planetary body. This discovery influenced the current debate within the astronomical community as to whether to treat tiny Pluto as a major planet or a large member of the collection of icy bodies that resides in the distant outer regions of the solar system. But, in August 2006, professional astronomers decided to delare Pluto to be a dwarf planet—despite its three moons. Human beings have learned more about the universe—including some of its strangest objects and most powerful phenomena—in the past four decades than in all of human history. The parade of new scientific discoveries from space-based astronomical observatories and scientific spacecraft is promoting an unprecedented rate of intellectual stimulation. It is not an unreasonable conclusion to suggest that space-based astronomy is influencing the trajectory of human civilization in ways never before imagined.

Chronology

··

✧ ca. 3000 B.C.E. (to perhaps 1000 B.C.E.)
Stonehenge erected on the Salisbury Plain of Southern England (possible use: ancient astronomical calendar for prediction of summer solstice)

✧ ca. 1300 B.C.E.
Egyptian astronomers recognize all the planets visible to the naked eye (Mercury, Venus, Mars, Jupiter, and Saturn), and they identify over 40 star patterns or constellations

✧ ca. 500 B.C.E.
Babylonians devise zodiac, which is later adopted and embellished by Greeks and used by other early peoples

✧ ca. 375 B.C.E.
The early Greek mathematician and astronomer Eudoxus of Cnidos starts codifying the ancient constellations from tales of Greek mythology

✧ ca. 275 B.C.E.
The Greek astronomer Aristarchus of Samos suggests an astronomical model of the universe (solar system) that anticipates the modern heliocentric theory proposed by Nicolaus Copernicus. However, these ideas, which Aristarchus presents in his work *On the Size and Distances of the Sun and the Moon,* are essentially ignored in favor of the geocentric model of the universe proposed by Eudoxus of Cnidus and endorsed by Aristotle

✧ ca. 129 B.C.E.
The Greek astronomer Hipparchus of Nicaea completes a catalog of 850 stars that remains important until the 17th century

✧ ca. 60 C.E.

The Greek engineer and mathematician Hero of Alexandria creates the aeoliphile, a toylike device that demonstrates the action-reaction principle that is the basis of operation of all rocket engines

✧ ca. 150 C.E.

Greek astronomer Ptolemy writes *Syntaxis* (later called the *Almagest* by Arab astronomers and scholars)—an important book that summarizes all the astronomical knowledge of the ancient astronomers, including the geocentric model of the universe that dominates Western science for more than one and a half millennia

✧ 820

Arab astronomers and mathematicians establish a school of astronomy in Baghdad and translate Ptolemy's work into Arabic, after which it became known as *al-Majisti* (The great work), or the *Almagest,* by medieval scholars

✧ 850

The Chinese begin to use gunpowder for festive fireworks, including a rocketlike device

✧ 1232

The Chinese army uses fire arrows (crude gunpowder rockets on long sticks) to repel Mongol invaders at the battle of Kaifung-fu. This is the first reported use of the rocket in warfare

✧ 1280–90

The Arab historian al-Hasan al-Rammah writes *The Book of Fighting on Horseback and War Strategies,* in which he gives instructions for making both gunpowder and rockets

✧ 1379

Rockets appear in western Europe; they are used in the siege of Chioggia (near Venice), Italy

✧ 1420

The Italian military engineer Joanes de Fontana writes *Book of War Machines,* a speculative work that suggests military applications of gunpowder rockets, including a rocket-propelled battering ram and a rocket-propelled torpedo

✧ 1429

The French army uses gunpowder rockets to defend the city of Orléans. During this period, arsenals throughout Europe begin to test various types of gunpowder rockets as an alternative to early cannons

✧ ca. 1500

According to early rocketry lore, a Chinese official named Wan-Hu attempted to use an innovative rocket-propelled kite assembly to fly through the air. As he sat in the pilot's chair, his servants lit the assembly's 47 gunpowder (black powder) rockets. Unfortunately, this early rocket test pilot disappeared in a bright flash and explosion

✧ 1543

The Polish church official and astronomer Nicolaus Copernicus changes history and initiates the Scientific Revolution with his book *De Revolutionibus Orbium Coelestium* (On the revolutions of the heavenly spheres). This important book, published while Copernicus lay on his deathbed, proposed a Sun-centered (heliocentric) model of the universe in contrast to the longstanding Earth-centered (geocentric) model advocated by Ptolemy and many of the early Greek astronomers

✧ 1608

The Dutch optician Hans Lippershey develops a crude telescope

✧ 1609

The German astronomer Johannes Kepler publishes *New Astronomy,* in which he modifies Nicolaus Copernicus's model of the universe by announcing that the planets have elliptical orbits rather than circular ones. Kepler's laws of planetary motion help put an end to more than 2,000 years of geocentric Greek astronomy

✧ 1610

On January 7, 1610, Galileo Galilei uses his telescope to gaze at Jupiter and discovers the giant planet's four major moons (Callisto, Europa, Io, and Ganymede). He proclaims this and other astronomical observations in his book, *Sidereus Nuncius* (Starry messenger). Discovery of these four Jovian moons encourages Galileo to advocate the heliocentric theory of Nicolaus Copernicus and brings him into direct conflict with church authorities

✧ 1642

Galileo Galilei dies while under house arrest near Florence, Italy, for his clashes with church authorities concerning the heliocentric theory of Nicolaus Copernicus

✧ **1647**

The Polish-German astronomer Johannes Hevelius publishes *Seleno-graphia*, in which he provides a detailed description of features on the surface (near side) of the Moon

✧ **1680**

Russian czar Peter the Great sets up a facility to manufacture rockets in Moscow. The facility later moves to St. Petersburg and provides the czarist army with a variety of gunpowder rockets for bombardment, signaling, and nocturnal battlefield illumination

✧ **1687**

Financed and encouraged by Sir Edmond Halley, Sir Isaac Newton publishes his great work, *Philosophiae Naturalis Principia Mathematica* (Mathematical principles of natural philosophy). This book provides the mathematical foundations for understanding the motion of almost everything in the universe including the orbital motion of planets and the trajectories of rocket-propelled vehicles

✧ **1780s**

The Indian ruler Hyder Ally (Ali) of Mysore creates a rocket corps within his army. Hyder's son, Tippo Sultan, successfully uses rockets against the British in a series of battles in India between 1782 and 1799

✧ **1804**

Sir William Congreve writes *A Concise Account of the Origin and Progress of the Rocket System* and documents the British military's experience in India. He then starts the development of a series of British military (black-powder) rockets

✧ **1807**

The British use about 25,000 of Sir William Congreve's improved military (black-powder) rockets to bombard Copenhagen, Denmark, during the Napoleonic Wars

✧ **1809**

The brilliant German mathematician, astronomer, and physicist Carl Friedrich Gauss publishes a major work on celestial mechanics that revolutionizes the calculation of perturbations in planetary orbits. His work paves the way for other 19th-century astronomers to mathematically anticipate and then discover Neptune (in 1846), using perturbations in the orbit of Uranus

✧ **1812**

British forces use Sir William Congreve's military rockets against American troops during the War of 1812. British rocket bombardment of Fort William McHenry inspires Francis Scott Key to add "the rocket's red glare" verse in the "Star Spangled Banner"

✧ **1865**

The French science fiction writer Jules Verne publishes his famous story *De la terre a la lune* (From the Earth to the Moon). This story interests many people in the concept of space travel, including young readers who go on to become the founders of astronautics: Robert Hutchings Goddard, Hermann J. Oberth, and Konstantin Eduardovich Tsiolkovsky

✧ **1869**

American clergyman and writer Edward Everett Hale publishes *The Brick Moon*—a story that is the first fictional account of a human-crewed space station

✧ **1877**

While a staff member at the U.S. Naval Observatory in Washington, D.C., the American astronomer Asaph Hall discovers and names the two tiny Martian moons, Deimos and Phobos

✧ **1897**

British author H. G. Wells writes the science fiction story *War of the Worlds*—the classic tale about extraterrestrial invaders from Mars

✧ **1903**

The Russian technical visionary Konstantin Eduardovich Tsiolkovsky becomes the first person to link the rocket and space travel when he publishes *Exploration of Space with Reactive Devices*

✧ **1918**

American physicist Robert Hutchings Goddard writes *The Ultimate Migration*—a far-reaching technology piece within which he postulates the use of an atomic-powered space ark to carry human beings away from a dying Sun. Fearing ridicule, however, Goddard hides the visionary manuscript; it remains unpublished until November 1972—many years after his death in 1945

✧ **1919**

American rocket pioneer Robert Hutchings Goddard publishes the Smithsonian monograph *A Method of Reaching Extreme Altitudes*. This important work presents all the fundamental principles of modern rocketry. Unfortunately, members of the press completely miss the true significance of his technical contribution and decide to sensationalize his comments about possibly reaching the Moon with a small, rocket-propelled package. For such "wild fantasy," newspaper reporters dubbed Goddard with the unflattering title of "Moon man"

✦ 1923

Independent of Robert Hutchings Goddard and Konstantin Eduardovich Tsiolkovsky, the German space-travel visionary Hermann J. Oberth publishes the inspiring book *Die Rakete zu den Planetenräumen* (The rocket into planetary space)

✦ 1924

The German engineer Walter Hohmann writes *Die Erreichbarkeit der Himmelskörper* (The attainability of celestial bodies)—an important work that details the mathematical principles of rocket and spacecraft motion. He includes a description of the most efficient (that is, minimum energy) orbit transfer path between two coplanar orbits—a frequently used space operations maneuver now called the Hohmann transfer orbit

✦ 1926

On March 16 in a snow-covered farm field in Auburn, Massachusetts, American physicist Robert Hutchings Goddard makes space technology history by successfully firing the world's first liquid-propellant rocket. Although his primitive gasoline (fuel) and liquid oxygen (oxidizer) device burned for only two and one half seconds and landed about 60 meters away, it represents the technical ancestor of all modern liquid-propellant rocket engines.

In April, the first issue of *Amazing Stories* appears. The publication becomes the world's first magazine dedicated exclusively to science fiction. Through science fact and fiction, the modern rocket and space travel become firmly connected. As a result of this union, the visionary dream for many people in the 1930s (and beyond) becomes that of interplanetary travel

✦ 1929

German space-travel visionary Hermann J. Oberth writes the award-winning book *Wege zur Raumschiffahrt* (Roads to space travel) that helps popularize the notion of space travel among nontechnical audiences

✦ 1933

P. E. Cleator founds the British Interplanetary Society (BIS), which becomes one of the world's most respected space-travel advocacy organizations

✧ 1935

Konstantin Tsiolkovsky publishes his last book, *On the Moon,* in which he strongly advocates the spaceship as the means of lunar and interplanetary travel

✧ 1936

P. E. Cleator, founder of the British Interplanetary Society, writes *Rockets through Space,* the first serious treatment of astronautics in the United Kingdom. However, several established British scientific publications ridicule his book as the premature speculation of an unscientific imagination

✧ 1939–1945

Throughout World War II, nations use rockets and guided missiles of all sizes and shapes in combat. Of these, the most significant with respect to space exploration is the development of the liquid propellant V-2 rocket by the German army at Peenemünde under Wernher von Braun

✧ 1942

On October 3, the German A-4 rocket (later renamed Vengeance Weapon Two or V-2 Rocket) completes its first successful flight from the Peenemünde test site on the Baltic Sea. This is the birth date of the modern military ballistic missile

✧ 1944

In September, the German army begins a ballistic missile offensive by launching hundreds of unstoppable V-2 rockets (each carrying a one-ton high explosive warhead) against London and southern England

✧ 1945

Recognizing the war was lost, the German rocket scientist Wernher von Braun and key members of his staff surrender to American forces near Reutte, Germany in early May. Within months, U.S. intelligence teams, under Operation Paperclip, interrogate German rocket personnel and sort through carloads of captured documents and equipment. Many of these German scientists and engineers join von Braun in the United States to continue their rocket work. Hundreds of captured V-2 rockets are also disassembled and shipped back to the United States.

On May 5, the Soviet army captures the German rocket facility at Peenemünde and hauls away any remaining equipment and personnel. In the closing days of the war in Europe, captured German rocket technology and personnel help set the stage for the great missile and space race of the cold war

On July 16, the United States explodes the world's first nuclear weapon. The test shot, code-named Trinity, occurs in a remote portion of southern New Mexico and changes the face of warfare forever. As part of the cold-war confrontation between the United States and the former Soviet Union, the nuclear-armed ballistic missile will become the most powerful weapon ever developed by the human race.

In October, a then-obscure British engineer and writer, Arthur C. Clarke, suggests the use of satellites at geostationary orbit to support global communications. His article in *Wireless World*, "Extra-Terrestrial Relays," represents the birth of the communications satellite concept—an application of space technology that actively supports the information revolution

✧ 1946

On April 16, the U.S. Army launches the first American-adapted, captured German V-2 rocket from the White Sands Proving Ground in southern New Mexico.

Between July and August the Russian rocket engineer Sergei Korolev develops a stretched-out version of the German V-2 rocket. As part of his engineering improvements, Korolev increases the rocket engine's thrust and lengthens the vehicle's propellant tanks

✧ 1947

On October 30, Russian rocket engineers successfully launch a modified German V-2 rocket from a desert launch site near a place called Kapustin Yar. This rocket impacts about 320 kilometers downrange from the launch site

✧ 1948

The September issue of the *Journal of the British Interplanetary Society* publishes the first in a series of four technical papers by L. R. Shepherd and A. V. Cleaver that explores the feasibility of applying nuclear energy to space travel, including the concepts of nuclear-electric propulsion and the nuclear rocket

✧ 1949

On August 29, the Soviet Union detonates its first nuclear weapon at a secret test site in the Kazakh Desert. Code-named First Lightning (Pervaya Molniya), the successful test breaks the nuclear-weapon monopoly enjoyed by the United States. It plunges the world into a massive nuclear arms race that includes the accelerated development of strategic ballistic missiles capable of traveling thousands of kilometers. Because they are well behind the United States in nuclear weapons technology, the leaders of the former Soviet Union decide to develop powerful, high-thrust rockets to carry their heavier, more primitive-design nuclear weapons. That decision gives the Soviet Union a major launch vehicle advantage when both superpowers decide to race into outer space (starting in 1957) as part of a global demonstration of national power

✧ 1950

On July 24, the United States successfully launches a modified German V-2 rocket with an American-designed WAC Corporal second-stage rocket from the U.S. Air Force's newly established Long Range Proving Ground at Cape Canaveral, Florida. The hybrid, multistage rocket (called Bumper 8) inaugurates the incredible sequence of military missile and space vehicle launches to take place from Cape Canaveral—the world's most famous launch site.

In November, British technical visionary Arthur C. Clarke publishes "Electromagnetic Launching as a Major Contribution to Space-Flight." Clarke's article suggests mining the Moon and launching the mined-lunar material into outer space with an electromagnetic catapult

✧ 1951

Cinema audiences are shocked by the science fiction movie *The Day the Earth Stood Still.* This classic story involves the arrival of a powerful, humanlike extraterrestrial and his robot companion, who come to warn the governments of the world about the foolish nature of their nuclear arms race. It is the first major science fiction story to portray powerful space aliens as friendly, intelligent creatures who come to help Earth.

Dutch-American astronomer Gerard Peter Kuiper suggests the existence of a large population of small, icy planetesimals beyond the orbit of Pluto—a collection of frozen celestial bodies now known as the Kuiper belt

✧ 1952

Collier's magazine helps stimulate a surge of American interest in space travel by publishing a beautifully illustrated series of technical articles written by space experts such as Wernher von Braun and Willey Ley. The first of the famous eight-part series appears on March 22 and is boldly titled "Man Will Conquer Space Soon." The magazine also hires the most influential space artist, Chesley Bonestell, to provide stunning color illustrations. Subsequent articles in the series introduce millions of American readers to the concept of a space station, a mission to the Moon, and an expedition to Mars

Wernher von Braun publishes *Das Marsprojekt* (The Mars project), the first serious technical study regarding a human-crewed expedition to

Mars. His visionary proposal involves a convoy of 10 spaceships with a total combined crew of 70 astronauts to explore the Red Planet for about one year and then return to Earth

✧ 1953

In August, the Soviet Union detonates its first thermonuclear weapon (a hydrogen bomb). This is a technological feat that intensifies the superpower nuclear arms race and increases emphasis on the emerging role of strategic, nuclear-armed ballistic missiles.

In October, the U.S. Air Force forms a special panel of experts, headed by John von Neumann, to evaluate the American strategic ballistic missile program. In 1954, this panel recommends a major reorganization of the American ballistic missile effort

✧ 1954

Following the recommendations of John von Neumann, President Dwight D. Eisenhower gives strategic ballistic missile development the highest national priority. The cold war missile race explodes on the world stage as the fear of a strategic ballistic missile gap sweeps through the American government. Cape Canaveral becomes the famous proving ground for such important ballistic missiles as the Thor, Atlas, Titan, Minuteman, and Polaris. Once developed, many of these powerful military ballistic missiles also serve the United States as space launch vehicles. U.S. Air Force General Bernard Schriever oversees the time-critical development of the Atlas ballistic missile—an astonishing feat of engineering and technical management

✧ 1955

Walt Disney (the American entertainment visionary) promotes space travel by producing an inspiring three-part television series that includes appearances by noted space experts like Wernher von Braun. The first episode, "Man in Space," airs on March 9 and popularizes the dream of space travel for millions of American television viewers. This show, along with its companion episodes, "Man and the Moon" and "Mars and Beyond," make von Braun and the term *rocket scientist* household words

✧ 1957

On October 4, Russian rocket scientist Sergei Korolev, with permission from Soviet premier Nikita S. Khrushchev, uses a powerful military rocket to successfully place *Sputnik 1* (the world's first artificial satellite) into orbit around Earth. News of the Soviet success sends a political and technical shockwave across the United States. The launch of *Sputnik 1* marks the beginning of the Space Age. It also is the start of the great space race of the cold war—a period when people measure national strength and global prestige by accomplishments (or failures) in outer space.

On November 3, the Soviet Union launches *Sputnik 2*—the world's second artificial satellite. It is a massive spacecraft (for the time) that carries a live dog named Laika, which is euthanized at the end of the mission.

The highly publicized attempt by the United States to launch its first satellite with a newly designed civilian rocket ends in complete disaster on December 6. The Vanguard rocket explodes after rising only a few inches above its launch pad at Cape Canaveral. Soviet successes with *Sputnik 1* and *Sputnik 2* and the dramatic failure of the Vanguard rocket heighten American anxiety. The exploration and use of outer space becomes a highly visible instrument of cold-war politics

✧ 1958

On January 31, the United States successfully launches *Explorer 1*—the first American satellite in orbit around Earth. A hastily formed team from the U.S. Army Ballistic Missile Agency (ABMA) and Caltech's Jet Propulsion Laboratory (JPL), led by Wernher von Braun, accomplishes what amounts to a national prestige rescue mission. The team uses a military ballistic missile as the launch vehicle. With instruments supplied by Dr. James Van Allen of the State University of Iowa, *Explorer 1* discovers Earth's trapped radiation belts—now called the Van Allen radiation belts in his honor.

The National Aeronautics and Space Administration (NASA) becomes the official civilian space agency for the United States government on October 1. On October 7, the newly created NASA announces the start of the Mercury Project—a pioneering program to put the first American astronauts into orbit around Earth.

In mid-December, an entire Atlas rocket lifts off from Cape Canaveral and goes into orbit around Earth. The missile's payload compartment carries Project Score (Signal Communications Orbit Relay Experiment)—a prerecorded Christmas season message from President Dwight D. Eisenhower. This is the first time the human voice is broadcast back to Earth from outer space

✧ 1959

On January 2, the Soviet Union sends a 790-pound- (360-kg-) mass spacecraft, *Lunik 1,* toward the Moon. Although it misses hitting the Moon by between 3,125 and 4,375 miles (5,000 and 7,000 km), it is the first human-made object to escape Earth's gravity and go in orbit around the Sun.

In mid-September, the Soviet Union launches *Lunik 2.* The 860-pound- (390-kg-) mass spacecraft successfully impacts on the Moon and becomes the first human-made object to (crash-) land on another world. *Lunik 2* carries Soviet emblems and banners to the lunar surface.

On October 4, the Soviet Union sends *Lunik 3* on a mission around the Moon. The spacecraft successfully circumnavigates the Moon and takes the first images of the lunar farside. Because of the synchronous rotation of the Moon around Earth, only the near side of the lunar surface is visible to observers on Earth

✧ 1960

The United States launches the *Pioneer 5* spacecraft on March 11 into orbit around the Sun. The modest-sized (92-pound- [42-kg-]) mass spherical American space probe reports conditions in interplanetary space between Earth and Venus over a distance of about 23 million miles (37 million km).

On May 24, the U.S. Air Force launches a MIDAS (Missile Defense Alarm System) satellite from Cape Canaveral. This event inaugurates an important American program of special military surveillance satellites intended to detect enemy missile launches by observing the characteristic infrared (heat) signature of a rocket's exhaust plume. Essentially unknown to the general public for decades because of the classified nature of their mission, the emerging family of missile surveillance satellites provides U.S. government authorities with a reliable early warning system concerning a surprise enemy (Soviet) ICBM attack. Surveillance satellites help support the national policy of strategic nuclear deterrence throughout the cold war and prevent an accidental nuclear conflict.

The U.S. Air Force successfully launches the *Discoverer 13* spacecraft from Vandenberg Air Force Base on August 10. This spacecraft is actually part of a highly classified Air Force and Central Intelligence Agency (CIA) reconnaissance satellite program called Corona. Started under special executive order from President Dwight D. Eisenhower, the joint agency spy satellite program begins to provide important photographic images of denied areas of the world from outer space. On August 18, *Discoverer 14* (also called *Corona XIV*) provides the U.S. intelligence community its first satellite-acquired images of the former Soviet Union. The era of satellite reconnaissance is born. Data collected by the spy satellites of the National Reconnaissance Office (NRO) contribute significantly to U.S. national security and help preserve global stability during many politically troubled times.

On August 12, NASA successfully launches the *Echo 1* experimental spacecraft. This large (100 foot [30.5 m] in diameter) inflatable, metalized balloon becomes the world's first passive communications satellite. At the dawn of space-based telecommunications, engineers bounce radio signals off the large inflated satellite between the United States and the United Kingdom.

The former Soviet Union launches *Sputnik 5* into orbit around Earth. This large spacecraft is actually a test vehicle for the new *Vostok* spacecraft that will soon carry cosmonauts into outer space. *Sputnik 5* carries two dogs, Strelka and Belka. When the spacecraft's recovery capsule functions properly the next day, these two dogs become the first living creatures to return to Earth successfully from an orbital flight

✧ 1961

On January 31, NASA launches a Redstone rocket with a Mercury Project space capsule on a suborbital flight from Cape Canaveral. The passenger astrochimp Ham is safely recovered down range in the Atlantic Ocean after reaching an altitude of 155 miles (250 km). This successful primate space mission is a key step in sending American astronauts safely into outer space.

The Soviet Union achieves a major space exploration milestone by successfully launching the first human being into orbit around Earth. Cosmonaut Yuri Gagarin travels into outer space in the *Vostok 1* spacecraft and becomes the first person to observe Earth directly from an orbiting space vehicle.

On May 5, NASA uses a Redstone rocket to send astronaut Alan B. Shepard, Jr., on his historic 15-minute suborbital flight into outer space from Cape Canaveral. Riding inside the Mercury Project *Freedom 7* space capsule, Shepard reaches an altitude of 115 miles (186 km) and becomes the first American to travel in space.

President John F. Kennedy addresses a joint session of the U.S. Congress on May 25. In an inspiring speech touching on many urgent national needs, the newly elected president creates a major space challenge for the United States when he declares: "I believe that this nation should commit itself to achieving the goal, before this decade is out, of landing a man on the Moon and returning him safely to Earth." Because of his visionary leadership, when American astronauts Neil A. Armstrong and Edwin E. "Buzz" Aldrin, Jr., step onto the lunar surface for the first time on July 20, 1969, the United States is recognized around the world as the undisputed winner of the cold-war space race

✧ 1962

On February 20, astronaut John Herschel Glenn, Jr., becomes the first American to orbit Earth in a spacecraft. An Atlas rocket launches the NASA Mercury Project *Friendship 7* space capsule from Cape Canaveral. After completing three orbits, Glenn's capsule safely splashes down in the Atlantic Ocean.

In late August, NASA sends the *Mariner 2* spacecraft to Venus from Cape Canaveral. *Mariner 2* passes within 21,700 miles (35,000 km) of the planet on December 14, 1962—thereby becoming the world's first successful interplanetary space probe. The spacecraft observes very high surface temperatures (~800°F [430°C]). These data shatter pre–space age visions about Venus being a lush, tropical planetary twin of Earth.

During October, the placement of nuclear-armed Soviet offensive ballistic missiles in Fidel Castro's Cuba precipitates the Cuban Missile Crisis. This dangerous superpower confrontation brings the world perilously close to nuclear warfare. Fortunately, the crisis dissolves when Premier Nikita S. Khrushchev withdraws the Soviet ballistic missiles after much skillful political maneuvering by President John F. Kennedy and his national security advisers

✧ 1964

On November 28, NASA's *Mariner 4* spacecraft departs Cape Canaveral on its historic journey as the first spacecraft from Earth to visit Mars. It successfully encounters the Red Planet on July 14, 1965, at a flyby distance of about 6,100 miles (9,800 km). *Mariner 4*'s close-up images reveal a barren, desertlike world and quickly dispel any pre–space age notions about the existence of ancient Martian cities or a giant network of artificial canals

✧ 1965

A Titan II rocket carries astronauts Virgil "Gus" I. Grissom and John W. Young into orbit on March 23 from Cape Canaveral, inside a two-person Gemini Project spacecraft. NASA's *Gemini 3* flight is the first crewed mission for the new spacecraft and marks the beginning of more sophisticated space activities by American crews in preparation for the Apollo Project lunar missions

✧ 1966

The former Soviet Union sends the *Luna 9* spacecraft to the Moon on January 31. The 220-pound- (100-kg-) mass spherical spacecraft soft lands in the Ocean of Storms region on February 3, rolls to a stop, opens four petal-like covers, and then transmits the first panoramic television images from the Moon's surface.

The former Soviet Union launches the *Luna 10* to the Moon on March 31. This massive (3,300-pound- [1,500-kg-] mass) spacecraft becomes the first human-made object to achieve orbit around the Moon.

On May 30, NASA sends the *Surveyor 1* lander spacecraft to the Moon. The versatile robot spacecraft successfully makes a soft landing (June 1) in the Ocean of Storms. It then transmits over 10,000 images from the lunar surface and performs numerous soil mechanics experiments in preparation for the Apollo Project human landing missions.

In mid-August, NASA sends the *Lunar Orbiter 1* spacecraft to the Moon from Cape Canaveral. It is the first of five successful missions to collect detailed images of the Moon from lunar orbit. At the end of each mapping mission, the orbiter spacecraft is intentionally crashed into the Moon to prevent interference with future orbital activities

✧ 1967

On January 27, disaster strikes NASA's Apollo Project. While inside their *Apollo 1* spacecraft during a training exercise on Launch Pad 34 at Cape Canaveral, astronauts Virgil "Gus" I. Grissom, Edward H. White, Jr., and Roger B. Chaffee are killed when a flash fire sweeps through their spacecraft. The Moon landing program is delayed by 18 months, while major design and safety changes are made in the Apollo Project spacecraft.

On April 23, tragedy also strikes the Russian space program when the Soviets launch cosmonaut Vladimir Komarov in the new *Soyuz* (union) spacecraft. Following an orbital mission plagued with difficulties, Komarov dies (on April 24) during reentry operations, when the spacecraft's parachute fails to deploy properly and the vehicle hits the ground at high speed

✧ 1968

On December 21, NASA's *Apollo 8* spacecraft (command and service modules only) departs Launch Complex 39 at the Kennedy Space Center during the first flight of the mighty Saturn V launch vehicle with a human crew as part of the payload. Astronauts Frank Borman, James Arthur Lovell, Jr., and William A. Anders become the first people to leave Earth's gravitational influence. They go into orbit around the Moon and capture images of an incredibly beautiful Earth "rising" above the starkly barren lunar horizon—pictures that inspire millions and stimulate an emerging environmental movement. After 10 orbits around the Moon, the first lunar astronauts return safely to Earth on December 27

✧ 1969

The entire world watches as NASA's *Apollo 11* mission leaves for the Moon on July 16 from the Kennedy Space Center. Astronauts Neil A. Armstrong, Michael Collins, and Edwin E. "Buzz" Aldrin, Jr., make a long-held dream of humanity a reality. On July 20, American astronaut Neil Armstrong cautiously descends the steps of the lunar excursion module's ladder and steps on the lunar surface, stating, "One small step for a man, one giant leap for mankind!" He and Buzz Aldrin become the first two people to walk on another world. Many people regard the Apollo Project lunar landings as the greatest technical accomplishment in all of human history

✧ 1970

NASA's *Apollo 13* mission leaves for the Moon on April 11. Suddenly, on April 13, a life-threatening explosion occurs in the service module portion of the Apollo spacecraft. Astronauts James A. Lovell, Jr., John Leonard Swigert, and Fred Wallace Haise, Jr., must use their lunar excursion module (LEM) as a lifeboat. While an anxious world waits and listens, the crew skillfully maneuvers their disabled spacecraft around the Moon. With critical supplies running low, they limp back to Earth on a free-return trajectory. At just the right moment on April 17, they abandon the LEM *Aquarius* and board the Apollo Project spacecraft (command module) for a successful atmospheric reentry and recovery in the Pacific Ocean

✧ 1971

On April 19, the former Soviet Union launches the first space station (called *Salyut 1*). It remains initially uncrewed because the three-cosmonaut crew of the *Soyuz 10* mission (launched on April 22) attempts to dock with the station but cannot go on board

✧ 1972

In early January, President Richard M. Nixon approves NASA's space shuttle program. This decision shapes the major portion of NASA's program for the next three decades.

On March 2, an Atlas-Centaur launch vehicle successfully sends NASA's *Pioneer 10* spacecraft from Cape Canaveral on its historic mission. This far-traveling robot spacecraft becomes the first to transit the main-belt asteroids, the first to encounter Jupiter (December 3, 1973) and by crossing the orbit of Neptune on June 13, 1983 (which at the time was the farthest planet from the Sun), the first human-made object ever to leave the planetary boundaries of the solar system. On an interstellar trajectory, *Pioneer 10* (and its twin, *Pioneer 11*) carries a special plaque, greeting any intelligent alien civilization that might find it drifting through interstellar space millions of years from now.

On December 7, NASA's *Apollo 17* mission, the last expedition to the Moon in the 20th century, departs from the Kennedy Space Center, propelled by a mighty Saturn V rocket. While astronaut Ronald E. Evans remains in lunar orbit, fellow astronauts Eugene A. Cernan and Harrison H. Schmitt become the 11th and 12th members of the exclusive Moon walkers club. Using a lunar rover, they explore the Taurus-Littrow region. Their safe return to Earth on December 19 brings to a close one of the epic periods of human exploration

✧ 1973

In early April, while propelled by Atlas-Centaur rocket, NASA's *Pioneer 11* spacecraft departs on an interplanetary journey from Cape Canaveral. The spacecraft encounters Jupiter (December 2, 1974) and then uses a gravity assist maneuver to establish a flyby trajectory to Saturn. It is the first spacecraft to view Saturn at close range (closest encounter on September 1, 1979) and then follows a path into interstellar space.

On May 14, NASA launches *Skylab*—the first American space station. A giant Saturn V rocket is used to place the entire large facility into orbit in a single launch. The first crew of three American astronauts arrives on May 25 and makes the emergency repairs necessary to save the station, which suffered damage during the launch ascent. Astronauts Charles (Pete) Conrad, Jr., Paul J. Weitz, and Joseph P. Kerwin stay onboard for 28 days. They are replaced by astronauts Alan L. Bean, Jack R. Lousma, and Owen K. Garriott, who arrive on July 28 and live in space for about 59 days. The final *Skylab* crew (astronauts Gerald P. Carr, William R. Pogue, and Edward G. Gibson) arrive on November 11 and reside in the station until February 8, 1974—setting a space endurance record (for the time) of 84 days. NASA then abandons *Skylab*.

In early November, NASA launches the *Mariner 10* spacecraft from Cape Canaveral. It encounters Venus (February 5, 1974) and uses a gravity assist maneuver to become the first spacecraft to investigate Mercury at close range

✧ 1975

In late August and early September, NASA launches the twin *Viking 1* (August 20) and *Viking 2* (September 9) orbiter/lander combination spacecraft to the Red Planet from Cape Canaveral. Arriving at Mars in 1976, all Viking Project spacecraft (two landers and two orbiters) perform exceptionally well—but the detailed search for microscopic alien life-forms on Mars remains inconclusive

✧ 1977

On August 20, NASA sends the *Voyager 2* spacecraft from Cape Canaveral on an epic grand tour mission, during which it encounters all four giant planets and then departs the solar system on an interstellar trajectory. Using the gravity assist maneuver, *Voyager 2* visits Jupiter (July 9, 1979), Saturn (August 25, 1981), Uranus (January 24, 1986), and Neptune (August 25, 1989). The resilient, far-traveling robot spacecraft (and its twin *Voyager 1*) also carries a special interstellar message from Earth—a digital record entitled *The Sounds of Earth*.

On September 5, NASA sends the *Voyager 1* spacecraft from Cape Canaveral on its fast trajectory journey to Jupiter (March 5, 1979), Saturn (March 12, 1980), and beyond the solar system

✧ 1978

In May, the British Interplanetary Society releases its Project Daedalus report—a conceptual study about a one-way robot spacecraft mission to Barnard's star at the end of the 21st century

✧ 1979

On December 24, the European Space Agency successfully launches the first Ariane 1 rocket from the Guiana Space Center in Kourou, French Guiana

✧ 1980

India's Space Research Organization successfully places a modest 77-pound-mass (35 kg) test satellite (called *Rohini*) into low Earth orbit on July 1. The launch vehicle is a four-stage, solid propellant rocket manufactured in India. The SLV-3 (Standard Launch Vehicle-3) gives India independent national access to outer space

✧ 1981

On April 12, NASA launches the space shuttle *Columbia* on its maiden orbital flight from Complex 39-A at the Kennedy Space Center. Astronauts John W. Young and Robert L. Crippen thoroughly test the new aerospace vehicle. Upon reentry, it becomes the first spacecraft to return to Earth by gliding through the atmosphere and landing like an airplane. Unlike all previous onetime use space vehicles, *Columbia* is prepared for another mission in outer space

✧ 1986

On January 24, NASA's *Voyager 2* spacecraft encounters Uranus.

On January 28, the space shuttle *Challenger* lifts off from the NASA Kennedy Space Center on its final voyage. At just under 74 seconds into the STS 51-L mission, a deadly explosion occurs, killing the crew and destroying the vehicle. Led by President Ronald Reagan, the United States mourns seven astronauts lost in the *Challenger* accident

✧ 1988

On September 19, the State of Israel uses a Shavit (comet) three-stage rocket to place the country's first satellite (called *Ofeq 1*) into an unusual east-to-west orbit—one that is opposite to the direction of Earth's rotation but necessary because of launch safety restrictions.

As the *Discovery* successfully lifts off on September 29 for the STS-26 mission, NASA returns the space shuttle to service following a 32-month hiatus after the *Challenger* accident

✧ 1989

On August 25, the *Voyager 2* spacecraft encounters Neptune

✧ 1994

In late January, a joint Department of Defense and NASA advanced technology demonstration spacecraft, *Clementine,* lifts off for the Moon from Vandenberg Air Force Base. Some of the spacecraft's data suggest that the Moon may actually possess significant quantities of water ice in its permanently shadowed polar regions

✧ 1995

In February, during NASA's STS-63 mission, the space shuttle *Discovery* approaches (encounters) the Russian *Mir* space station as a prelude to the development of the *International Space Station.* Astronaut Eileen Marie Collins serves as the first female shuttle pilot.

On March 14, the Russians launch the *Soyuz TM-21* spacecraft to the *Mir* space station from the Baikanour Cosmodrome. The crew of three includes American astronaut Norman Thagard—the first American to travel into outer space on a Russian rocket and the first to stay on the *Mir* space station. The *Soyuz TM-21* cosmonauts also relieve the previous *Mir* crew, including cosmonaut Valeri Polyakov, who returns to Earth on March 22 after setting a world record for remaining in space for 438 days.

In late June, NASA's space shuttle *Atlantis* docks with the Russian *Mir* space station for the first time. During this shuttle mission (STS-71), *Atlantis* delivers the *Mir 19* crew (cosmonauts Anatoly Solovyev and Nikolai Budarin) to the Russian space station and then returns the *Mir 18* crew back to Earth—including American astronaut Norman Thagard, who has just spent 115 days in space onboard the *Mir.* The Shuttle-*Mir* docking program is the first phase of the *International Space Station.* A total of nine shuttle-*Mir* docking missions will occur between 1995 and 1998

✧ 1998

In early January, NASA sends the *Lunar Prospector* to the Moon from Cape Canaveral. Data from this orbiter spacecraft reinforce previous hints that the Moon's polar regions may contain large reserves of water ice in a mixture of frozen dust lying at the frigid bottom of some permanently shadowed craters.

In early December, the space shuttle *Endeavour* ascends from the NASA Kennedy Space Center on the first assembly mission of the *International Space Station*. During the STS-88 shuttle mission, *Endeavour* performs a rendezvous with the previously launched Russian-built *Zarya* (sunrise) module. An international crew connects this module with the American-built *Unity* module carried in the shuttle's cargo bay

✧ 1999

In July, astronaut Eileen Marie Collins serves as the first female space shuttle commander (STS-93 mission) as the *Columbia* carries NASA's *Chandra X-ray Observatory* into orbit

✧ 2001

NASA launches the *Mars Odyssey 2001* mission to the Red Planet in early April—the spacecraft successfully orbits the planet in October

✧ 2002

On May 4, NASA successfully launches its *Aqua* satellite from Vandenberg Air Force Base. This sophisticated Earth-observing spacecraft joins the *Terra* spacecraft in performing Earth system science studies.

On October 1, the United States Department of Defense forms the U.S. Strategic Command (USSTRATCOM) as the control center for all American strategic (nuclear) forces. USSTRATCOM also conducts military space operations, strategic warning and intelligence assessment, and global strategic planning

✧ 2003

On February 1, while gliding back to Earth after a successful 16-day scientific research mission (STS-107), the space shuttle *Columbia* experiences a catastrophic reentry accident at an altitude of about 63 km over the Western United States. Traveling at 18 times the speed of sound, the orbiter vehicle disintegrates, taking the lives of all seven crew members: six American astronauts (Rick Husband, William McCool, Michael Anderson, Kalpana Chawla, Laurel Clark, and David Brown) and the first Israeli astronaut (Ilan Ramon).

NASA's Mars Exploration Rover (MER) *Spirit* is launched by a Delta II rocket to the Red Planet on June 10. *Spirit,* also known as MER-A, arrives safely on Mars on January 3, 2004, and begins its teleoperated surface exploration mission under the supervision of mission controllers at the NASA Jet Propulsion Laboratory.

NASA launches the second Mars Exploration Rover, called *Opportunity,* using a Delta II rocket launch, which lifts off from Cape Canaveral Air Force Station on July 7, 2003. *Opportunity,* also called MER-B, successfully lands on Mars on January 24, 2004, and starts its teleoperated surface exploration mission under the supervision of mission controllers at the NASA Jet Propulsion Laboratory

✧ 2004

On July 1, NASA's *Cassini* spacecraft arrives at Saturn and begins its four-year mission of detailed scientific investigation.

In mid-October, the Expedition 10 crew, riding a Russian launch vehicle from Baikonur Cosmodrome, arrives at the *International Space Station* and the Expedition 9 crew returns safely to Earth.

On December 24, the 703-pound- (319-kg-) mass *Huygens* probe successfully separates from the *Cassini* spacecraft and begins its journey to Saturn's moon, Titan

✧ 2005

On January 14, the *Huygens* probe enters the atmosphere of Titan and successfully reaches the surface some 147 minutes later. *Huygens* is the first spacecraft to land on a moon in the outer solar system.

On July 4, NASA's Deep Impact mission successfully encounters Comet Tempel 1.

NASA successfully launches the space shuttle *Discovery* on the STS-114 mission on July 26 from the Kennedy Space Center in Florida. After docking with the *International Space Station*, the *Discovery* returns to Earth and lands at Edwards AFB, California, on August 9.

On August 12, NASA launches the *Mars Reconnaissance Orbiter* from Cape Canaveral AFS, Florida.

On September 19, NASA announces plans for a new spacecraft designed to carry four astronauts to the Moon and to deliver crews and supplies to the *International Space Station*. NASA also introduces two new shuttle-derived launch vehicles: a crew-carrying rocket and a cargo-carrying, heavy-lift rocket.

The *Expedition 12* crew (Commander William McArthur and Flight Engineer Valery Tokarev) arrives at the *International Space Station* on October 3 and replaces the *Expedition 11* crew.

The People's Republic of China successfully launches its second human spaceflight mission, called *Shenzhou 6*, on October 12. Two taikonauts, Fei Junlong and Nie Haisheng, travel in space for almost five days and make 76 orbits of Earth before returning safely to Earth, making a soft, parachute-assisted landing in northern Inner Mongolia

✧ 2006

On January 15, the sample package from NASA's *Stardust* spacecraft, containing comet samples, successfully returns to Earth.

NASA launches the *New Horizons* spacecraft from Cape Canaveral on January 19 and successfully sends this robot probe on its long one-way mission to conduct a scientific encounter with the Pluto system (in 2015) and then to explore portions of the Kuiper belt that lie beyond.

Follow-up observations by NASA's *Hubble Space Telescope*, reported on February 22, confirm the presence of two new moons around the distant planet Pluto. The moons, tentatively called S/2005 P 1 and S/2005 P 2, were first discovered by *Hubble* in May 2005, but the science team wants to examine the Pluto system further to characterize the orbits of the new moons and validate the discovery.

NASA scientists announce on March 9 that the *Cassini* spacecraft may have found evidence of liquid water reservoirs that erupt in Yellowstone Park–like geysers on Saturn's moon Enceladus.

On March 10, NASA's *Mars Reconnaissance Orbiter* successfully arrives at Mars and begins a six-month-long process of adjusting and trimming the shape of its orbit around the Red Planet prior to performing its operational mapping mission.

The *Expedition 13* crew (Commander Pavel Vinogradov and Flight Engineer Jeff Williams) arrive at the *International Space Station* on April 1 and replace the *Expedition 12* crew. Joining them for several days before returning back to Earth with the *Expedition 12* crew is Brazil's first astronaut, Marcos Pontes

On August 24, the members of the International Astronomical Union (IAU) meet for the organization's 2006 General Assembly in Prague, Czech Republic. After much heated debate, the 2,500 assembled professional astronomers decide (by vote) to demote Pluto from its traditional status as one of the nine major planets and place the object into a new class, called a dwarf planet. The IAU decision now leaves the solar system with eight major planets and three dwarf planets: Pluto (which serves as the prototype dwarf planet), Ceres (the largest asteroid), and the large, distant Kuiper belt object indentified as 2003 UB313 (nicknamed "Xena"). Astronomers anticipate the discovery of other dwarf planets in the distant parts of the solar system.

Glossary

absolute magnitude (symbol: M)   The measure of the brightness (or apparent magnitude) that a star would have if it were hypothetically located at a reference distance of 10 parsecs (pc), about 32.6 light-years, from the Sun.

absolute temperature   A temperature value relative to absolute zero, which corresponds to 0 K. In almost all modern scientific activities, absolute temperature values are expressed in kelvins (K)—a unit within the international system (SI) unit named in honor of the Scottish physicist Baron William Thomson Kelvin (1827–1907).

absorption line   The gap, dip, or dark line feature in a stellar spectrum occurring at a specific wavelength. It is caused by the absorption of the radiation emitted from a star's hotter interior regions by an absorbing substance in its relatively cooler outer regions. Analysis of absorption lines lets astronomers determine the chemical composition of stars.

accretion disk   The whirling disk of inflowing (or infalling) material from a normal stellar companion that develops around a massive compact object, such as a neutron star or a black hole. The conservation of angular momentum shapes this disk, which is often accompanied by a pair of very-high-speed material jets departing in opposite directions perpendicular to the plane of the disk.

acronym   A word formed from the first letters of a name, such as *HST*, which stands for the *Hubble Space Telescope*. It is also a word formed by combining the initial parts of a series of words, such as *lidar*—which means *l*ight *d*etection *a*nd *r*anging. Acronyms are frequently used in space technology and astronomy.

active galaxies   Collectively, those unusual celestial objects, including quasars, bl lac objects, and Seyfert galaxies, that have extremely energetic central regions called active galactic nuclei (AGN). These emit enormous amounts of electromagnetic radiation, ranging from radio waves to X-rays and gamma rays.

albedo   The ratio of the amount of electromagnetic radiation (such as visible light) reflected by a surface to the total amount of electromagnetic radiation incident upon the surface. The albedo is usually expressed as a percentage. For example, the planetary albedo of Earth is about 30 percent.

Alpha Centauri   The closest star system, about 4.3 light-years away. It is actually a triple star system, with two stars orbiting around each other and a third star, called Proxima Centauri, revolving around the pair at some distance.

alphanumeric (*alphabet* plus *numeric*)   Including letters and numerical digits; for example, JEN75WX11.

Amor group   A collection of near-Earth asteroids that crosses the orbit of Mars but does not cross the orbit of Earth. This asteroid group acquired its name from the 0.6-mile- (1-km-) diameter Amor asteroid, discovered in 1932 by the Belgian astronomer Eugène Joseph Delporte (1882–1995).

ancient constellations   The collection of approximately 50 constellations drawn up by ancient astronomers and recorded by the Greek astronomer Ptolemy (ca. 100 C.E.–ca. 170 C.E.), including such familiar constellations as the signs of the zodiac: Ursa Major (The Great Bear), Boötes (The Herdsman), and Orion (The Hunter).

Andromeda Galaxy   The Great Spiral Galaxy (or M31) in the constellation of Andromeda, about 2.2 million light-years way. It is the most distant object visible to the naked eye and is the closest spiral galaxy to the Milky Way Galaxy.

angstrom (symbol: Å)   A unit of length often used in physics to quantify the wavelength of radiation in the visible, near-infrared, and near-ultraviolet portions of the electromagnetic spectrum. Named after the Swedish physicist Anders Jonas Ångstrom (1814–74). One angstrom equals 0.1 nanometers.

angular measure   Units of angle generally expressed in terms of degrees (°), arc minutes ('), and arc seconds ("), where one degree of angle equals 60 arc minutes, and one arc minute equals 60 arc seconds.

antenna   A device used to detect, collect, or transmit radio waves. A radio telescope is a large receiving antenna. Many spacecraft have both a directional antenna and an omnidirectional antenna to transmit (downlink) telemetry and to receive (uplink) instructions.

antimatter   Matter in which the ordinary nuclear particles (such as electrons, protons, and neutrons) are replaced by their corresponding antiparticles—that is, positrons, antiprotons, antineutrons, and so on. It is sometimes called mirror matter. Normal matter and antimatter mutually annihilate each other upon contact and are converted into pure energy called annihilation radiation.

apastron   The point in a body's orbit around a star at which it is at a maximum distance from the star. *Compare with* PERIASTRON.

aphelion   The point in an object's orbit around the Sun that is most distant from the Sun. *Compare with* PERIHELION.

Aphrodite Terra   A large, fractured highland region near the equator of Venus.

apogee   The point in the orbit of a satellite that is farthest from Earth. The term applies to both the orbit of the Moon as well as to the orbits of artificial satellites around Earth. At apogee the orbital velocity of a satellite is at a minimum. *Compare with* PERIGEE.

Apollo group   A collection of near-Earth asteroids that have perihelion distances of 1.017 astronomical units (AU) or less, taking them across the orbit of Earth around the Sun. This group acquired its name from the asteroid Apollo—the first member to be discovered, in 1932, by the German astronomer Karl Reinmuth (1892–1979).

apolune   That point in an orbit around the Moon of a spacecraft launched from the lunar surface that is farthest from the Moon. *Compare with* PERILUNE.

apparent   In astronomy, observed. True values are reduced from apparent (observed) values by eliminating those factors, such as refraction and flight time, which can affect the observation.

apparent magnitude (symbol: m)   The brightness of a star (or other celestial body) as measured by an observer on Earth. Its value depends on the star's intrinsic brightness (luminosity), how far away it is, and how much of its light has been absorbed by the intervening interstellar medium.

apparent motion   The observed motion of a heavenly body across the celestial sphere, assuming that Earth is at the center of the celestial sphere and is standing still (stationary).

Arecibo Observatory   The world's largest radio/radar telescope, with a 1,000-foot- (305-m-) diameter dish. It is located in a large, bowl-shaped natural depression in the tropical jungles of Puerto Rico.

asteroid   A small, solid rocky object that orbits the Sun but is independent of any major planet. Most asteroids (or minor planets) are found in the main asteroid belt between the orbits of Mars and Jupiter. The largest asteroid is Ceres, which was discovered in 1801 by Italian astronomer Giuseppe Piazzi (1746–1826). Earth-crossing asteroids, or near-Earth asteroids (NEAs), have orbits that take them near or across Earth's orbit around the Sun and are divided into the Aten, Apollo, and Amor groups.

astro-   A prefix that means "star" or (by extension) "outer space" or "celestial"; for example, the words *astronaut*, *astronautics*, and *astrophysics*.

astrometric binary   A binary star system in which irregularities in the proper motion (wobbling) of a visible star imply the presence of an undetected companion.

astrometry   Branch of astronomy that involves the very precise measurement of the motion and position of celestial bodies.

astronomical unit (AU)   A convenient unit of distance defined as the semimajor axis of Earth's orbit around the Sun. One AU, the average distance between Earth and the Sun, is equal to approximately 93×10^6 miles (149.6×10^6 km), or 499.01 light-seconds.

Aten group   A collection of near-Earth asteroids that crosses the orbit of Earth, but whose average distances from the Sun lie inside Earth's orbit. This asteroid group acquired its name from the 0.55-mile- (0.9-km-) diameter asteroid Aten, discovered in 1976 by the American astronomer Eleanor Kay Helin (née Francis).

atmosphere   The gravitationally bound gaseous envelope that forms an outer region around a planet or other celestial body.

atmospheric probe   The special collection of scientific instruments (usually released by a mother spacecraft) for determining the pressure, composition, and temperature of a planet's atmosphere at different altitudes. An example is the probe released by NASA's *Galileo* spacecraft in December 1995. As it plunged into Jupiter's atmosphere, the probe successfully transmitted its scientific data to *Galileo* (the mother spacecraft) for about 58 minutes.

atmospheric window   A wavelength interval within which a planet's atmosphere is transparent (that is, easily transmits) electromagnetic radiation.

Barnard's star   A red dwarf star approximately six light-years from the Sun, making it the fourth-nearest star to the solar system. Discovered in 1916 by American astronomer Edward Emerson Barnard (1857–1923), it has the largest proper motion (some 10.3 seconds of arc per year) of any known star.

barred spiral galaxy   A type of spiral galaxy that has a bright bar of stars across the central regions of the galactic nucleus.

basin (impact)   A large, shallow lowland area in the crust of a terrestrial planet formed by the impact of an asteroid or comet.

big bang (theory)   A contemporary theory in cosmology concerning the origin of the universe. It suggests that a very large, ancient explosion started space and time of the present universe, which has been expanding ever since.

big crunch   Within the closed universe model of cosmology, the postulated end state that occurs after the present universe expands to its maximum physical dimensions and then collapses in on itself under the influence of gravitation, eventually reaching an infinitely dense end point, or singularity.

binary star system   A pair of stars that orbits around a common center of mass and is bound together by mutual gravitation.

blackbody   A perfect emitter and perfect absorber of electromagnetic radiation. According to Max Planck's radiation law, the radiant energy emitted by a blackbody is a function only of the absolute temperature of the emitting object.

black dwarf   The cold remains of a white dwarf star that no longer emits visible radiation, or a nonradiating ball of interstellar gas that has contracted under gravitation but contains too little mass to initiate nuclear fusion.

black hole   An incredibly compact, gravitationally collapsed mass from which nothing (light, matter, or any other kind of information) can escape. Astrophysicists believe that a black hole is the natural end product when a very massive star dies and collapses beyond a certain critical dimension.

blazar   A variable extragalactic object (possibly a high-speed jet from an active galactic nucleus) that exhibits very dynamic, sometimes violent behavior.

bl lac (or bl lacertae) object   A class of extragalactic objects thought to be the active centers of faint elliptical galaxies that vary considerably in brightness over very short periods of time (typically hours, days, or weeks). Scientists further speculate that a very-high-speed (relativistic) jet is emerging from such an object straight at an observer on Earth.

blue giant   A massive, very-high-luminosity star with a surface temperature of about 30,000 K that has exhausted all its hydrogen thermonuclear fuel and left the main sequence.

blueshift   *See* DOPPLER SHIFT.

brown dwarf   A very-low-luminosity, substellar (almost a star) celestial object that contains starlike material (that is, hydrogen and helium) but has too low a mass (typically 1 to 10 percent of a solar mass) to allow its core to initiate thermonuclear fusion (hydrogen burning).

caldera   A large volcanic depression that formed by one of three basic geologic processes: explosion, collapse, or erosion.

Callisto   *See* GALILEAN SATELLITES.

Caloris basin   A very large, ringed, impact basin on Mercury.

Cassini mission   The joint NASA–European Space Agency planetary exploration mission to Saturn launched from Cape Canaveral on October 15, 1997. Since July 2004, the *Cassini* spacecraft has performed detailed studies of Saturn, its rings, and its moons. The *Cassini* mother spacecraft also delivered the *Huygens* probe, which successfully plunged into the nitrogen-rich atmosphere of Titan (Saturn's largest moon) on January 14,

2005. The mother spacecraft is named after the Italian-French astronomer Giovanni Cassini (1625–1712); the Titan probe is named after Dutch astronomer Christiaan Huygens (1629–95).

celestial body   A heavenly body. Any aggregation of matter in outer space constituting a unit for study in astronomy, such as planets, moons, comets, asteroids, stars, nebulae, and galaxies.

Cepheid variable   A type of very bright, supergiant star that exhibits a regular pattern of changing its brightness as a function of time. The period of this pulsation pattern is directly related to the star's intrinsic brightness. Modern astronomers use Cepheid variables to determine astronomical distances.

Ceres   The first and largest (580-mile- [940-km-] diameter) asteroid to be found. It was discovered on January 1, 1801, by Italian astronomer Giuseppe Piazzi (1746–1826).

***Chandra X-ray Observatory* (*CXO*)**   One of NASA's major orbiting astronomical observatories launched in July 1999 and named after the Indian-American astrophysicist Subrahmanyan Chandrasekhar (1910–95). Previously called the *Advanced X-ray Astrophysics Facility*. This Earth-orbiting facility studies some of the most interesting and puzzling X-ray sources in the universe, including emissions from active galactic nuclei, exploding stars, neutron stars, and matter falling into black holes.

charge coupled device (CCD)   An electronic (solid-state) device containing a regular array of sensor elements that are sensitive to various types of electromagnetic radiation (e.g., light) and emit electrons when exposed to such radiation. The emitted electrons are collected and the resulting charge analyzed. CCDs are used as the light-detecting component in modern television cameras and telescopes.

Charon   The large (about 745-mile- [1,200-km-] diameter) moon of Pluto discovered in 1978 by the American astronomer James Walter Christy.

chasma   A canyon or deep linear feature on a planet's surface.

chromatic aberration   A phenomenon that occurs in a refracting optical system because light of different wavelengths (colors) is refracted (bent) by a different amount. As a result, a simple lens will give red light a longer focal length than blue light.

chromosphere   The reddish layer in the Sun's atmosphere located between the photosphere (the apparent solar surface) and the base of the corona. It is the source of solar prominences.

Chryse Planitia   A large plain on Mars characterized by many ancient channels that could have once contained flowing surface water. It was the landing site for NASA's *Viking 1* lander (robot spacecraft) in July 1976.

cislunar   Of or pertaining to phenomena, projects, or activities happening in the region of outer space between Earth and the Moon. From the Latin word *cis*, meaning "on this side," and the word *lunar*, which means "of or pertaining to the Moon." Therefore, it means "on this side of the Moon."

closed universe   The model in cosmology that assumes the total mass of the universe is sufficiently large that one day the galaxies will slow down and stop expanding because of their mutual gravitational attraction. At that time, the universe will have reached its maximum size, and then gravitation will make it slowly contract, ultimately collapsing to a single point of infinite density (sometimes called the big crunch). Also called bounded universe model. *Compare with* OPEN UNIVERSE.

cluster of galaxies   An accumulation of galaxies that lie within a few million light-years of each other and are bound by gravitation. Galactic clusters can occur with just a few member galaxies (say 10 to 100), such as the Local Group, or they can occur in great groupings involving thousands of galaxies.

cold war   The ideological conflict between the United States and the former Soviet Union that lasted from approximately 1946 to 1989, involving rivalry, mistrust, and hostility just short of overt military action. The tearing down of the Berlin Wall in November 1989 is generally considered as the symbolic end of the cold-war period.

coma   The gaseous envelope that surrounds the nucleus of a comet.

comet   A dirty ice-"rock" consisting of dust, frozen water, and gases that orbits the Sun. As a comet approaches the inner solar system from deep space, solar radiation causes its frozen materials to vaporize (sublime), creating a coma and a long tail of dust and ions. Scientists think these icy planetesimals are the remainders of the primordial material from which the outer planets were formed billions of years ago. *See also* KUIPER BELT and OORT CLOUD.

Comet Halley (1P/Halley)   The most famous periodic comet. Named after British astronomer Edmond Halley (1656–1742), who correctly predicted its 1758 return. Reported since 240 B.C.E., this comet reaches perihelion approximately every 76 years. During its most recent inner solar-system appearance, an international fleet of five different robot spacecraft, including the *Giotto* spacecraft, performed scientific investigations that supported the dirty ice-rock model of a comet's nucleus.

compact body   A small, very dense celestial body that represents the end product of stellar evolution: a white dwarf, a neutron star, or a black hole.

Compton Gamma Ray Observatory (*CGRO*)   A major NASA orbiting astrophysical observatory dedicated to gamma-ray astronomy. The CGRO was placed in orbit around Earth in April 1991. At the end of its useful scientific mission, flight controllers intentionally commanded the massive (35,900-pound- [16,300-kg-] mass) spacecraft to perform a de-orbit burn. This caused it to reenter and safely crash in June 2000 in a remote region of the Pacific Ocean. The spacecraft was named in honor of the American physicist Arthur Holly Compton (1892–1962).

constellation   1. (aerospace) A term used to describe collectively the number and orbital disposition of a set of satellites. 2. (astronomy) An easily identifiable configuration of the brightest stars in a moderately small region of the night sky.

continuously habitable zone (CHZ)   The region around a star in which one or several planets (or possibly their moons) can maintain conditions appropriate for the emergence and sustained existence of life. One important characteristic of a planet in the CHZ is that its environmental conditions support the retention of significant amounts of liquid water on the planetary surface.

corona   The outermost region of a star. The Sun's corona consists of low-density clouds of very hot gases (> 1 million K) and ionized materials.

Cosmic Background Explorer (*COBE*)   A NASA robot spacecraft placed in orbit around Earth in November 1989. It successfully measured the spectrum and intensity distribution of the cosmic microwave background (CMB).

cosmic microwave background (CMB)   The background of microwave radiation that permeates the universe and has a blackbody temperature of about 2.7 K. Sometimes called the primal glow, scientists believe it represents the remains of the ancient fireball in which the universe was created.

cosmic ray   Extremely energetic particle (usually a bare atomic nucleus) that moves through outer space at speeds just below the speed of light and bombards Earth from all directions. Galactic cosmic rays are samples of material from outside the solar system and provide direct evidence of phenomena that occur as a result of explosive processes in stars throughout the Milky Way Galaxy. Solar cosmic rays (mostly protons and alpha particles) are ejected from the Sun during solar flare events.

cosmic-ray astronomy   The branch of high-energy astrophysics that uses cosmic rays to provide information on the origin of the chemical elements through nucleosynthesis during stellar explosions.

cosmological principle   The hypothesis that the expanding universe is isotropic and homogeneous. In other words, there is no special location for observing the universe and all observers anywhere in the universe would see the same recession of distant galaxies.

cosmology   The study of the origin, evolution, and structure of the universe. Contemporary cosmology centers around the big bang hypothesis—a theory stating that about 14 billion years ago the universe began in a great explosion and has been expanding ever since.

Crab nebula   The supernova remnant of an exploding star observed in 1054 by Chinese astronomers. It is about 6,500 light-years away in the Constellation Taurus and contains a pulsar that flashes optically.

crater   A bowl-shaped topographic depression with steep slopes on the surface of a planet or moon. There are two general types: impact (as formed by an asteroid, comet, or meteoroid strike) and eruptive (as formed when a volcano erupts).

crust   The outermost solid layer of a planet or moon.

Cygnus X-1   The strong X-ray source in the constellation Cygnus that scientists believe comes from a binary star system, consisting of an orbiting supergiant star and a black-hole companion. Gas drawn off the supergiant star emits X-rays as it is intensely heated while falling into the black hole.

dark matter   Matter in the universe that cannot be observed directly because it emits very little or no electromagnetic radiation. Scientists infer its existence through secondary phenomena such as gravitational effects and suggest that it may make up about 90 percent of the total mass of the universe. Also called missing mass.

dark nebula   A cloud of interstellar dust and gas sufficiently dense and thick that the light from more distant stars and celestial bodies (behind it) is obscured.

Deep Space Network (DSN)   NASA's global network of antennae that serves as the radio-wave communications link to distant interplanetary spacecraft and probes, transmitting instructions to them and receiving data from them. Large radio antennae of the DSN's three Deep Space Communications Complexes are located in Goldstone, California; near Madrid, Spain; and near Canberra, Australia. It provides almost continuous contact with a spacecraft in deep space as Earth rotates on its axis.

deep-space probe   A spacecraft designed for exploring deep space, especially to the vicinity of the Moon and beyond. This includes lunar probes, Mars probes, outer planet probes, solar probes, and so on.

degenerate star   A star that has collapsed to a high-density condition, such as a white dwarf or a neutron star.

Deimos   The tiny (about 7.5- miles- [12-km-] diameter), irregularly shaped outer moon of Mars, discovered in 1877 by the American astronomer Asaph Hall (1829–1907).

Doppler shift   The apparent change in the observed frequency and wavelength of a source due to the relative motion of the source and an observer. If the source is approaching the observer, the observed frequency is higher and the observed wavelength is shorter. This change to shorter wavelengths is often called the blueshift. If the source is moving away from the observer, the observed frequency will be lower and the wavelength will be longer. This change to longer wavelengths is called the redshift. Named after Austrian physicist Christian Johann Doppler (1803–53), who discovered this physical phenomenon in 1842 by observing sound waves.

downlink   The telemetry signal received at a ground station from a spacecraft or space probe.

dwarf galaxy   A small, often-elliptical galaxy containing a million (106) to perhaps a billion (109) stars. The Magellanic Clouds, humans' nearest galactic neighbors, are examples.

dwarf planet   As defined by the International Astronomical Union (IAU) in August 2006, a celestial body that (a) is in orbit around the Sun, (b) has sufficient mass for its self-gravity to overcome rigid body forces so that it assumes a nearly round shape, (c) has not cleared the cosmic neighborhood around its orbit, and (d) is not a satellite or another (larger) body. Included in this definition are Pluto, Ceres (the largest asteroid), and 2003 UB313 (a large, distant Kuiper belt object, nicknamed "Xena").

dwarf star   Any star that is a main-sequence star, according to the Hertzsprung-Russell diagram. Most stars found in the galaxy, including the Sun, are of this type and are from 0.1 to about 100 solar masses in size. However, when astronomers use the term *dwarf star*, they are not referring to white dwarfs, brown dwarfs, or black dwarfs, which are celestial bodies that are not in the collection of main-sequence stars.

Earth-crossing asteroid   *See* ASTEROID.

ecliptic (plane)   The apparent annual path of the Sun among the stars; the intersection of the plane of Earth's orbit around the Sun with the celestial sphere. Because of the tilt in Earth's axis, the ecliptic is a great circle of the celestial sphere inclined at an angle of about 23.5 degrees to the celestial equator.

electromagnetic radiation (EMR)   Radiation composed of oscillating electric and magnetic fields and propagated with the speed of light. EMR includes (in order of decreasing energy and increasing wavelength) gamma rays, X-rays, ultraviolet radiation, visible radiation (light), infrared radiation, radar waves, and radio waves.

elliptical galaxy   A galaxy with a smooth, elliptical shape without spiral arms and having little or no interstellar gas and dust.

elliptical orbit   A noncircular, Keplerian orbit. *See also* KEPLER'S LAWS.

encounter   The close flyby or rendezvous of a spacecraft with a target body. The target of an encounter can be a natural celestial body (such as a planet, asteroid, or comet) or a human-made object (such as another spacecraft).

escape velocity (V_e)   The minimum velocity that an object must acquire to overcome the gravitational attraction of a celestial body. The escape velocity for an object launched from the surface of Earth is approximately seven miles (11.2 km) per second, while the escape velocity from the surface of Mars is about three miles (5.0 km) per second.

Europa   *See* GALILEAN SATELLITES.

European Space Agency (ESA)   An international organization that promotes the peaceful use of outer space and cooperation among the European member states in space research and applications.

event horizon   The point of no return for a black hole; the distance from a black hole within which nothing can escape. Also called the Schwarzschild radius.

evolved star   A star near the end of its lifetime when most of its hydrogen fuel has been exhausted; a star that has left the main sequence.

expanding universe   Any model of the universe in modern cosmology that has the distance between widely separated celestial objects (e.g., distant galaxies) continuing to grow or expand with time.

exploding galaxies   Violent, very energetic explosions centered in certain galactic nuclei where the total mass of ejected material is comparable to the mass of some 5 million, average-size, Sun-like stars. Jets of gas 1,000 light-years long are also typical.

Explorer 1   The first U.S. Earth-orbiting satellite, which was launched successfully from Cape Canaveral on January 31, 1958 (local time), by a Juno I four-stage configuration of the Jupiter C launch vehicle.

Explorer spacecraft   NASA has used the name "Explorer" to designate members of a large family of scientific robot spacecraft and satellites intended to "explore the unknown." Since 1958, Explorer spacecraft have studied Earth's atmosphere and ionosphere; the planet's precise shape and geophysical features; the planet's magnetosphere and interplanetary space; and various astronomical and astrophysical phenomena.

extragalactic   Occurring, located, or originating beyond the Milky Way Galaxy.

extragalactic astronomy   A branch of astronomy that started about 1930 and deals with everything in the universe outside of the Milky Way Galaxy.

extrasolar   Occurring, located, or originating outside of the solar system.

extrasolar planet   A planet around a star other than the Sun.

***Extreme Ultraviolet Explorer* (*EUVE*)**   The 70th NASA Explorer spacecraft. After being successfully launched from Cape Canaveral in June 1992, this scientific robot spacecraft went into orbit around Earth and provided astronomers with a survey of the (until then) relatively unexplored extreme-ultraviolet portion of the electromagnetic spectrum.

facula   A bright region of the Sun's photosphere.

farside   The side of the Moon that never faces Earth.

fixed stars   A term used by early astronomers to distinguish between the apparently motionless background stars and the wandering stars (planets). Modern astronomers now use this term to describe stars that have no detectable proper motion.

flare (solar)   A bright eruption from the Sun's corona.

flyby   An interplanetary or deep-space mission in which the flyby spacecraft passes close to its target celestial body (e.g., a distant planet, moon, asteroid, or comet) but does not impact the target or go into orbit around it.

fossa   A long, narrow, shallow (ditch-like) depression found on the surface of a planet or a moon.

frequency (usual symbol: f or ν)   The rate of repetition of a recurring or regular event; the number of cycles of a wave per second. For electromagnetic radiation, the frequency (ν) is equal to the speed of light (c) divided by the wavelength *See also* HERTZ.

fusion (nuclear)   The nuclear process by which lighter atomic nuclei join (or fuse) to create a heavier nucleus. Thermonuclear reactions are fusion reactions caused by very high temperatures (millions of kelvins). The energy of the Sun and other stars comes from thermonuclear fusion reactions.

galactic cluster   A diffuse collection of from 10 to perhaps several hundred stars, loosely held together by the force of gravitation. The term *open cluster* is now preferred by astronomers.

galactic nucleus   The central region of a galaxy.

galaxy   A very large accumulation of stars with from 1 million (10^6) to a million million (10^{12}) members. These island universes come in a variety of sizes and shapes, from dwarf galaxies (like the Magellanic Clouds) to majestic spiral galaxies (like the Andromeda Galaxy). Astronomers classify them as elliptical, spiral (or barred spiral), or irregular.

Galilean satellites   The four largest and brightest moons of Jupiter, discovered in 1610 by the Italian astronomer Galileo Galilei (1564–1642). They are Io, Europa, Ganymede, and Callisto.

Galileo Project   NASA's highly successful scientific mission to Jupiter launched in October 1989. With electricity supplied by two radioisotope-thermoelectric generator units, the *Galileo* spacecraft extensively studied the Jovian system from December 1995 until February 2003. Upon arrival it also released a probe into the upper portions of Jupiter's atmosphere. On February 28, 2003, the NASA flight team terminated its operation of the *Galileo* spacecraft and commanded the robot craft to plunge into Jupiter's atmosphere. This mission-ending plunge took place in late September 2003.

gamma ray (symbol: γ)   Very-short-wavelength, high-frequency packets (or quanta) of electromagnetic radiation. Gamma-ray photons are similar to X-rays, except that they originate within the atomic nucleus and have energies between 10,000 electron volts (10 keV) and 10 million electron volts (10 MeV).

gamma-ray astronomy   Branch of astronomy based on the detection of the energetic gamma rays associated with supernovas, exploding galaxies, quasars, pulsars, and phenomena near suspected black holes.

Ganymede   *See* GALILEAN SATELLITES.

geocentric   Relative to Earth as the center; measured from the center of Earth.

giant-impact model   The hypothesis that the Moon originated when a Mars-sized object struck a young Earth with a glancing blow. The giant (oblique) impact released material that formed an accretion disk around Earth, out of which the Moon formed.

giant molecular cloud (GMC)   Massive clouds of gas in interstellar space composed primarily of molecules of hydrogen and dust. GMCs can contain enough mass to make several million stars like the Sun and are often the sites of star formation.

giant planets   In this solar system, the large, gaseous outer planets: Jupiter, Saturn, Uranus, and Neptune. Any detected or suspected extrasolar planets as large or larger than Jupiter.

giant star   A star near the end of its life that has swollen in size, such as a blue giant or a red giant.

***Giotto* spacecraft**   Scientific robot spacecraft launched by the European Space Agency in July 1985 that successfully encountered the nucleus of Comet Halley in mid-March 1986 at a (closest approach) distance of about 370 miles (600 km).

globular cluster   Compact cluster of up to 1 million, generally older, stars.

gravitation   The force of attraction between two masses. From Sir Isaac Newton's law of gravitation, this attractive force operates along a line joining the centers of mass, and its magnitude is inversely proportional to the square of the distance between the two masses. From Albert Einstein's general relativity theory, gravitation is viewed as a distortion of the space-time continuum.

gravitational collapse   The unimpeded contraction of any mass caused by its own gravity.

gravity   The attraction of a celestial body for any nearby mass. Specifically, the downward force imparted by Earth on a mass near or on its surface.

gravity assist   The change in a spacecraft's direction and speed achieved by a carefully calculated flyby through a planet's gravitational field. This change in spacecraft velocity occurs without the use of supplementary propulsive energy.

Greenwich mean time   Mean solar time at the meridian of Greenwich, England, used as the basis for standard time throughout the world. Normally expressed in four numerals, 0001 to 2400. Also called universal time.

H I region   A diffuse region of neutral, predominantly atomic hydrogen in interstellar space.

H II region   A region in interstellar space consisting mainly of ionized hydrogen and existing mostly in discrete clouds.

Halley's comet   *See* COMET HALLEY.

halo orbit   A circular or elliptical orbit in which a spacecraft remains in the vicinity of a Lagrangian libration point.

hard landing   A relatively high-velocity impact of a lander spacecraft or probe on a solid planetary surface. The impact usually destroys all equipment, except perhaps a very rugged instrument package or payload container.

heliocentric   With the Sun as a center.

heliopause   The boundary of the heliosphere. It is thought to occur about 100 astronomical units from the Sun and marks the edge of the Sun's influence and the beginning of interstellar space.

heliosphere   The region of outer space within the boundary of the heliopause in which the solar wind flows. Contains the Sun and the solar system.

hertz (symbol: Hz)   The SI unit of frequency. One hertz is equal to one cycle per second. Named in honor of German physicist Heinrich Rudolf Hertz (1857–94), who produced and detected radio waves for the first time in 1888.

Hertzspung-Russell (H-R) diagram   A useful graphic depiction of the different types of stars arranged according to their spectral classification and luminosity. Named in honor of the Danish astronomer Ejnar Hertzsprung (1873–1967) and the American astronomer Henry Norris Russell (1877–1957), who developed the diagram independently of one another.

high Earth orbit (HEO)   An orbit around Earth at an altitude greater than 3,475 miles (5,600 km).

High-Energy Astronomy Observatory (HEAO)   A series of three NASA robot spacecraft placed in Earth's orbit (*HEAO-1* launched in August 1977; *HEAO-2* in November 1978; and *HEAO-3* in September 1979) to support X-ray astronomy and gamma-ray astronomy. After launch, NASA renamed *HEAO-2* the *Einstein Observatory* to honor the famous German-Swiss-American physicist Albert Einstein (1879–1955).

highlands   Oldest-exposed areas on the surface of the Moon; extensively cratered and chemically distinct from the maria.

Hohmann transfer orbit   The most efficient orbit transfer path between two coplanar circular orbits. The maneuver consists of two impulsive high-thrust burns (or firings) of a spacecraft's propulsion system. The technique was suggested in 1925 by the German engineer Walter Hohmann (1880–1945).

housekeeping   The collection of routine tasks that must be performed to keep a spacecraft functioning properly during an orbital flight or interplanetary mission.

Hubble Space Telescope (HST)   A cooperative European Space Agency and NASA program to operate a long-lived, space-based optical observatory. Launched on April 25, 1990, by NASA's space shuttle *Discovery* (STS-31 mission), subsequent on-orbit repair and refurbishment missions have allowed this powerful Earth-orbiting optical observatory to revolutionize knowledge of the size, structure, and makeup of the universe. Named in honor of the American astronomer Edwin Powell Hubble (1889–1953).

Huygens **probe**   A scientific probe sponsored by the European Space Agency and named after the Dutch astronomer Christiaan Huygens (1629–95). The *Cassini* mother spacecraft delivered *Huygens* to Saturn, and the probe successfully plunged into the nitrogen-rich atmosphere of Titan (Saturn's largest moon) on January 14, 2005.

hyperbolic orbit   An orbit in the shape of a hyperbola; all interplanetary, flyby spacecraft follow hyperbolic orbits, both for Earth departure and again upon arrival at the target planet.

Imbrium basin   Large (about 810 miles [1,300 km] across), ancient impact crater on the Moon.

impact crater *See* CRATER.

in-flight phase The flight of a robot spacecraft from launch to the time of planetary flyby, encounter and orbit, or impact.

infrared astronomy The branch of astronomy dealing with infrared radiation from relatively cool celestial objects, such as interstellar clouds of dust and gas (typically 100 K) and stars with surface temperatures below about 6,000 K.

infrared radiation That portion of the electromagnetic spectrum between the optical (visible) and radio wavelengths. The infrared region extends from about one micrometer (µm) to 1,000-µm wavelength.

inner planets The terrestrial planets: Mercury, Venus, Earth, and Mars—all of which have orbits around the Sun that lie inside the main asteroid belt. *Compare with* OUTER PLANETS.

interferometer An instrument that achieves high angular resolution by combining signals from at least two widely separated telescopes (optical interferometer) or a widely separated antenna array (radio interferometer).

intergalactic Between or among the galaxies.

interplanetary Between the planets; within the solar system.

interplanetary dust Tiny particles of matter (generally less than 100 micrometers in diameter) that exist in outer space within the confines of this solar system.

interstellar Between or among the stars.

interstellar medium The gas and tiny dust particles that are found between the stars in the Milky Way Galaxy. Over 100 different types of molecules have been discovered in interstellar space, including many organic molecules.

interstellar probe A conceptual, highly automated robot spacecraft launched by human beings in this solar system (or perhaps by intelligent alien beings in some other solar system) to explore nearby star systems.

Io *See* GALILEAN SATELLITES.

ionizing radiation Any type of atomic or nuclear radiation that displaces electrons from atoms or molecules, thereby producing ions within the irradiated material. Examples include: alpha (α) radiation, beta (β) radiation, gamma (γ) radiation, protons, neutrons, and X-rays.

irregular galaxy A galaxy with a poorly defined structure or shape.

Ishtar Terra A very large highland plateau in the northern hemisphere of Venus, about 3,100 miles (5,000 km) long and 370 miles (600 km) wide.

island universe Term coined in the 18th century by German philosopher Immanuel Kant (1724–1804) to describe distant collections of stars—the term that replaced it is *galaxy*.

Jovian planet A large (Jupiter-like) planet characterized by a great total mass, low average density, mostly liquid interior, and an abundance of lighter elements (especially hydrogen and helium). In this solar system, the Jovian planets are Jupiter, Saturn, Uranus, and Neptune.

Kepler's laws The three empirical laws describing the motion of a satellite (natural or human-made) in orbit around its primary body, formulated in the early 17th century by the German astronomer Johannes Kepler (1571–1630).

Kuiper belt A region in the outer solar system beyond Neptune and extending out to perhaps 1,000 astronomical units that contains millions of icy planetesimals. These icy objects range in size from tiny particles to Pluto-sized planetary bodies. The Dutch-American astronomer Gerard Peter Kuiper (1905–73) first suggested the existence of this disk-shaped reservoir of icy objects in 1951. *See also* OORT CLOUD.

Lagrangian libration point The five points in outer space (called L_1, L_2, L_3, L_4, and L_5) where a small object can experience a stable orbit in spite of the force of gravity exerted by two much more massive celestial bodies when they orbit about a common center of mass. Joseph-Louis Lagrange calculated the existence and location of these points in 1772.

lander (spacecraft) A spacecraft designed to safely reach the surface of a planet or moon and survive long enough on the planetary body to collect useful scientific data that it sends back to Earth by telemetry.

Large Magellanic Cloud (LMC)   An irregular galaxy about 20,000 light-years in diameter and approximately 160,000 light-years from Earth. *See also* MAGELLANIC CLOUDS.

light   The portion of the electromagnetic spectrum that can be seen by the human eye. Visible light (radiation) ranges from approximately 750 nanometers (long-wavelength, red) to about 370 nanometers (short-wavelength, violet).

light-year (symbol: ly)   The distance light (or other forms of electromagnetic radiation) travels in one year. One light-year equals a distance of approximately 5.87×10^{12} miles (9.46×10^{12} km), or 63,240 astronomical units (AU).

limb   The visible outer edge or observable rim of the disk of a celestial body.

Local Group   A small cluster of about 30 galaxies, of which the Milky Way Galaxy and the Andromeda Galaxy are dominant members.

long-period comet   A comet with an orbital period around the Sun greater than 200 years. *Compare with* SHORT-PERIOD COMET.

low Earth orbit (LEO)   A circular orbit just above Earth's sensible atmosphere at an altitude of between 185 and 250 miles (300 and 400 km).

luminosity (symbol: L)   The rate at which a star or other luminous object emits energy, usually in the form of electromagnetic radiation.

Luna   A series of robot spacecraft sent to explore the Moon in the 1960s and 1970s by the former Soviet Union.

lunar   Of or pertaining to Earth's natural satellite, the Moon.

lunar crater   A depression, usually circular, on the surface of the Moon. *See also* CRATER.

lunar highlands   The light-colored, heavily-cratered mountainous part of the Moon's surface.

lunar orbiter   A spacecraft placed in orbit around the Moon; specifically, the series of five *Lunar Orbiter* robot spacecraft NASA used from 1966 to

1967 to photograph the Moon's surface precisely in support of the Apollo Project.

lunar probe   A planetary probe for exploring and reporting conditions on or about the Moon.

Lunar Prospector   A NASA orbiter spacecraft that circled the Moon from 1998 to 1999, searching for mineral resources. Data collected by this robot spacecraft suggest the possible presence of water-ice deposits in the Moon's permanently shadowed polar regions.

lunar rover   Crewed or automated (robot) rover vehicles used to explore the Moon's surface. NASA's lunar rover vehicle served as a Moon car for Apollo Project astronauts during the *Apollo 15, 16,* and *17* expeditions. Russian *Lunokhod 1* and *2* robot rovers were operated on the Moon from Earth between 1970 and 1973.

Lunokhod   A Russian eight-wheeled robot vehicle, controlled by radio-wave signals from Earth and used to perform lunar surface exploration during the *Luna 17* (1970) and *Luna 21* (1973) missions to the Moon.

Magellanic Clouds   The two dwarf, irregularly shaped neighboring galaxies that are closest to our Milky Way Galaxy. The Large Magellanic Cloud is about 160,000 light-years away and the Small Magellanic Cloud approximately 180,000 light-years away. Both can be seen with the naked eye in the Southern Hemisphere, and their presence was first recorded in 1519 by the Portuguese explorer Ferdinand Magellan (1480–1521), after whom they are named.

Magellan mission   The planetary orbiter spacecraft that used its powerful radar-imaging system to make detailed surface maps of cloud-covered Venus from 1990 to 1994. NASA named this robot spacecraft *Magellan,* after the Portuguese explorer Ferdinand Magellan (1480–1521).

magnetosphere   The region around a planet in which charged atomic particles are influenced (and often trapped) by the planet's own magnetic field rather than the magnetic field of the Sun as projected by the solar wind.

magnitude   A number, measured on a logarithmic scale, that indicates the relative brightness of a celestial object. The smaller the magnitude number, the greater the brightness. Ancient astronomers called the brightest stars of the night sky stars of the first magnitude, because they were the first visible after sunset. Other stars were called 2nd, 3rd, 4th, 5th, and 6th magnitude stars according to their relative brightness. Sixth magnitude stars are the faintest stars visible to the naked eye. In 1856 the British astronomer Norman Robert Pogson (1829–91) proposed a more precise logarithmic magnitude in which a difference of five magnitudes represents a relative brightness ratio of 100 to 1, while a difference of one magnitude is 2.512. This scale is now widely used in modern astronomy.

main belt asteroid   One located in the asteroid belt between Mars and Jupiter.

main-sequence star   A star in the prime of its life that shines with a constant luminosity by steadily converting hydrogen into helium through thermonuclear fusion in its core.

maria (singular: mare)   Latin word for "seas." Originally used by the Italian astronomer Galileo Galilei (1564–1642) to describe the large, dark ancient lava flows on the lunar surface, since he and other 17th-century astronomers thought these features were bodies of water on the Moon's surface. Following tradition, this term is still used by modern astronomers.

Mariner   A series of NASA planetary exploration robot spacecraft that performed important flyby and orbital missions to Mercury, Mars, and Venus in the 1960s and 1970s.

Mars Exploration Rovers (MERs)   In 2003 NASA launched identical twin Mars rovers designed to operate on the surface of the Red Planet. *Spirit* (MER-A) was launched from Cape Canaveral on June 10, 2003, and successfully landed on Mars on January 4, 2004. *Opportunity* (MER-B) was launched from Cape Canaveral on July 7, 2003, and successfully landed on Mars on January 25, 2004. Both soft landings used the airbag bounce-and-roll arrival demonstrated during the Mars Pathfinder mission. *Spirit* landed in Gusev Crater, and *Opportunity* landed at Terra Meridiania. As of September 1, 2006, both rovers were still functioning.

Mars Global Surveyor   A NASA orbiter spacecraft launched in November 1996 that has performed detailed studies of the Martian surface and atmosphere since March 1999.

Mars Odyssey   Launched from Cape Canaveral by NASA in April 2001, the *2001 Mars Odyssey* is an orbiter spacecraft designed to conduct a detailed exploration of Mars, with emphasis being given to the search for geological features that would indicate the presence of water—flowing on the surface in the past or currently frozen in subsurface reservoirs.

Mars Pathfinder   An innovative NASA mission that successfully landed a Mars surface rover—a small robot called *Sojourner*—in the Ares Vallis region of the Red Planet in July 1997. For over 80 days, human beings on Earth used teleoperation and telepresence to cautiously drive the six-wheeled mini-rover to interesting locations on the Martian surface.

Maxwell Montes   A prominent mountain range on Venus located in Ishtar Terra, containing the highest peak (6.8-mile- [11-km-] altitude) on the planet. The mountain range was named after the Scottish theoretical physicist James Clerk Maxwell (1831–79).

megaparsec (Mpc)   One million parsecs; a distance of approximately 3,260,000 light-years.

Milky Way Galaxy   Humans' home galaxy—a large spiral galaxy that contains between 200 and 600 billion solar masses. The Sun lies some 30,000 light-years from the galactic center.

minor planet   *See* ASTEROID.

missing mass   *See* DARK MATTER.

moon   A small, natural celestial body that orbits a larger one; a natural satellite.

Moon   Earth's only natural satellite and closest celestial neighbor. It has an equatorial diameter of 2,159 miles (3,476 km), keeps the same side (nearside) toward Earth, and orbits at an average distance (center to center) of 238,758 miles (384,400 km).

mother spacecraft   A exploration spacecraft that carries and deploys one or several atmospheric probes, lander spacecraft, and/or lander and rover spacecraft combinations when it arrives at a target planet. The mother spacecraft then relays data back to Earth and may also orbit the planet in order to perform its own scientific mission. NASA's *Galileo* spacecraft to Jupiter and *Cassini* spacecraft to Saturn are examples.

multispectral sensing   The remote-sensing method of simultaneously collecting several different bands (wavelength regions) of electromagnetic radiation (such as the visible, the near-infrared, and the thermal infrared bands) when observing an object or region of interest.

nadir   The direction from a spacecraft directly down toward the center of a planet. It is the opposite of the ZENITH.

naked eye   The normal human eye unaided by any optical instrument, such as a telescope. The use of corrective lens (glasses) or contact lens that restore an individual's normal vision falls under the category of naked-eye observing.

NASA   The National Aeronautics and Space Administration, the civilian space agency of the United States. Created in 1958 by an act of Congress, NASA's overall mission is to plan, direct, and conduct civilian (including scientific) aeronautical and space activities for peaceful purposes.

near-Earth asteroid (NEA)   *See* ASTEROID.

nearside   The side of the Moon that always faces Earth.

nebula (plural: nebulae or nebulas)   A cloud of interstellar gas or dust. It can be seen as either a dark hole against a brighter background (called a dark nebula) or as a luminous patch of light (called a bright nebula).

New Horizons Pluto-Kuiper belt Flyby mission   NASA's reconnaissance-type exploration mission that will help scientists understand the icy worlds at the outer edge of the solar system. Successfully launched from Cape Canaveral on January 19, 2006, the *New Horizons* spacecraft is currently scheduled to perform a flyby of Pluto and its moon (Charon) in 2015. The robot spacecraft will then continue beyond Pluto and visit one or more Kuiper belt objects (of opportunity) by 2026. The robot spacecraft's long journey will help resolve some basic questions about the surface features and properties of these distant icy bodies as well as their geology, interior makeup, and atmospheres.

nova (plural: novas or novae)   From the Latin for "new," a highly evolved star that exhibits a sudden and exceptional brightness, usually temporary, and then returns to its former luminosity. A nova is now thought to be the outburst of a degenerate star in a binary star system.

nucleosynthesis   The production of heavier chemical elements from the fusion (joining together) of lighter chemical elements (such as hydrogen and helium) in thermonuclear reactions in the interior of stars.

nucleus   1. (cometary) The small (few miles diameter), permanent, solid ice-rock central body of a comet. 2. (galactic) The central region of a galaxy, a few light-years in diameter, where matter is concentrated.

observable universe   The portions of the universe that can be detected and studied by the light they emit.

observatory   The place (or facility) from which astronomical observations are made. For example, the Keck Observatory is a ground-based observatory, while the *Hubble Space Telescope* is a space-based (or Earth-orbiting) observatory.

Olympus Mons   A huge mountain on Mars about 405 miles (650 km) wide and rising 16 miles (26 km) above the surrounding plains—the largest-known shield volcano in the solar system.

Oort cloud   The large number (about 10^{12}) or cloud of comets postulated in 1950 by the Dutch astronomer Jan Hendrik Oort (1900–92) to orbit the Sun at an enormous distance—ranging from some 50,000 and 80,000 astronomical units.

open universe   A universe that will continue to expand forever. *See* COSMOLOGY.

orbit   The path followed by a body in space, generally under the influence of gravity—as, for example, a satellite around a planet.

orbiter   A robot spacecraft especially designed to travel through interplanetary space, achieve a stable orbit around the target planet (or other celestial body), and conduct a program of detailed scientific investigation.

Orbiting Astronomical Observatory   A series of large, Earth-orbiting astronomical observatories developed by NASA in the 1960s to broaden humans' understanding of the universe, especially as related to ultraviolet astronomy.

oscillating universe   A closed universe in which gravitational collapse is followed by a new wave of expansion. *See* COSMOLOGY.

outer planets   The major planets in this solar system with orbits greater than the orbit of Mars, including Jupiter, Saturn, Uranus, and Neptune, Pluto is now a dwarf planet.

parking orbit   The temporary (but stable) orbit of a spacecraft around a celestial body, used for assembly and/or transfer of equipment or to wait for conditions favorable for departure from that orbit.

parsec (symbol: pc)   An astronomical unit of distance corresponding to a trigonometric parallax (π) of one second of arc. The term is a shortened form of *parallax sec*ond, and one parsec represents a distance of 3.26 light-years (or 206,265 astronomical units).

perfect cosmological principle   The postulation that at all times the universe appears the same to all observers. *See also* COSMOLOGY.

peri-   A prefix meaning "near."

periastron   The point of closest approach of two stars in a binary star system. *Compare with* APASTRON.

perigee   The point at which a satellite's orbit is the closest to its primary body; the minimum altitude attained by an Earth-orbiting object. *Compare with* APOGEE.

perihelion   The point in an elliptical orbit around the Sun that is nearest to the center of the Sun. *Compare with* APHELION.

perilune   The point in an elliptical orbit around the Moon that is nearest to the lunar surface. *Compare with* APOLUNE.

periodic comet   A comet with a period of less than 200 years. Also called a short-period comet.

Phobos   The larger, innermost of the two small moons of Mars, discovered in 1877 by the American astronomer Asaph Hall (1829–1907). *See also* DEIMOS.

photometer   An instrument that measures light intensity and the brightness of celestial objects, such as stars.

photosphere   The intensely bright (white-light), visible surface of the Sun or other star.

Pioneer 10, 11 **spacecraft**   NASA's twin robot spacecraft that were the first to navigate the main asteroid belt, the first to visit Jupiter (1973 and 1974), the first to visit Saturn (*Pioneer 11* in 1979), and the first human-made objects to leave the solar system (*Pioneer 10* in 1983). Each space-craft is now on a different trajectory to the stars, carrying a special message (the Pioneer plaque) for any intelligent alien civilization that might find it millions of years from now.

Pioneer Venus mission   Two spacecraft launched by NASA to Venus in 1978. *Pioneer 12* was an orbiter spacecraft that gathered data from 1978 to 1992. The *Pioneer Venus Multiprobe* served as a mother spacecraft, launching one large and three identical small planetary probes into Venus's atmosphere (December 1978).

plage   A bright patch in the Sun's chromosphere.

Planck's radiation law   The physical principle, developed in 1900 by the German physicist Max Karl Planck (1858–1947). This physical principle describes the distribution of energy radiated by a blackbody. With this law, Planck introduced his concept of the quantum (or photon) as a small unit of energy responsible for the transfer of electromagnetic radiation.

planet   A nonluminous celestial body that orbits around the Sun or some other star. The name *planet* is from the ancient Greek *planetes* ("wanderers")—since early astronomers identified the planets as the wandering points of light relative to the fixed stars. There are eight major planets, three dwarf planets, and numerous minor planets (or asteroids) in humans' solar system. In August 2006, the International Astronomical Union (IAU) clarified the difference between a planet and a dwarf planet. A planet is defined as a celestial body that (a) is in orbit around the Sun, (b) has sufficient mass for its self-gravity to overcome rigid body forces so that it assumes a nearly round shape, and (c) has cleared the cosmic neighborhood around its orbit. There are now eight major planets in the solar system: Mercury, Venus, Earth, Mars, Jupiter, Saturn, Uranus, and Neptune. Pluto is now a dwarf planet. *See also* DWARF PLANET.

planetary nebula   The shell of gas ejected from the outer layers of an extremely hot star (like a red giant) at the end of its life cycle.

planetary probe   An instrument-containing robot spacecraft deployed in the atmosphere or on the surface of a planetary body in order to obtain environmental information.

planetesimals Small rock and rock/ice celestial objects found in the solar system, ranging from 0.06 mile (0.1 km) to about 62 miles (100 km) in diameter.

planet fall The act of landing a spacecraft or space vehicle on a planet or moon.

planetoid *See* ASTEROID.

polar orbit An orbit around a planet (or primary body) that passes over or near its poles; an orbit with an inclination of about 90 degrees.

Population I stars Hot, luminous, young stars, including those like the Sun, that reside in the disk of a spiral galaxy and are higher in heavy element content (about 2 percent abundance) than Population II stars.

Population II stars Older stars that are lower in heavy element content than Population I stars and reside in globular clusters as well as in the halo of a galaxy—that is, the distant spherical region that surrounds a galaxy.

primary body The celestial body around which a satellite, moon, or other object orbits or from which it is escaping or toward which it is falling.

prominence A cloud of cooler plasma extending high above the Sun's visible surface, rising above the photosphere into the corona.

proper motion (symbol: μ) The apparent angular displacement of a star with respect to the celestial sphere.

protogalaxy A galaxy in the early stages of evolution.

protoplanet Any of a star's planets as such planets emerge during the process of accretion in which planetesimals collide and coalesce into large objects.

protostar A star in the making. Specifically, the stage in a young star's evolution after it has separated from a gas cloud but prior to its collapsing sufficiently (due to gravity) to support thermonuclear fusion reactions in its core.

Proxima Centauri The closest star to the Sun—the third member of the Alpha Centauri triple-star system. It is some 4.2 light-years away.

pulsar A rapidly spinning neutron star that generates regular pulses of electromagnetic radiation. Although originally discovered by radio-wave observations, pulsars have since been observed at optical, X-ray, and gamma-ray energies.

Quaoar Large, icy world with a diameter of about 780 miles (1,250 km) located in the Kuiper belt about 1 million miles (1.6 million km) beyond Pluto. It was first observed in June 2004.

quasar A mysterious, very distant object with a high redshift—that is, traveling away from Earth at great speed. These objects appear almost like stars but are far more distant than any individual star now observed. They might be the very luminous centers of active distant galaxies. When first identified in 1963, they were called *quasi*-stell*ar* radio sources—or quasars. Also called quasi-stellar object.

radiation belt The region(s) in a planet's magnetosphere where there is a high density of trapped atomic particles from the solar wind.

radio frequency (RF) The portion of the electromagnetic spectrum useful for telecommunications with a frequency range between 10,000 and 3 $\times 10^{11}$ hertz.

radio galaxy A galaxy (often dumbbell-shaped) that produces very strong radio-wave signals. Cygnus A is an example of intense source about 650 million light-years away.

Ranger Project The first NASA robot spacecraft sent to the Moon in the 1960s. These hard-impact planetary probes were designed to take a series of television images of lunar surface before crash-landing.

rays (lunar) Bright streaks extending across the surface from young impact craters on the Moon; also observed on Mercury and on several large moons of the outer planets.

red dwarf (star) Reddish main-sequence stars (spectral type K and M) that are relatively cool (about 4000 K surface temperature) and have low mass (about 0.5 solar mass or less). These faint, low-luminosity stars are inconspicuous, yet they represent the most common type in universe and the longest-lived. Barnard's star is an example.

red giant (star)   A large, cool star with a surface temperature of about 2,500 K and a diameter 10 to 100 times that of the Sun. This type of highly luminous star is at the end of its evolutionary life, departing the main sequence after exhausting the hydrogen in its core. It is often a variable star. Some 5 billion years from now the Sun will evolve into a massive red giant.

Red Planet   The planet Mars—so named because of its distinctive reddish soil.

redshift   *See* DOPPLER SHIFT.

reflecting telescope   *See* TELESCOPE.

refracting telescope   *See* TELESCOPE.

regolith (lunar)   The unconsolidated mass of surface debris that overlies the Moon's bedrock. This blanket of pulverized lunar dust and soil was created by millions of years of meteoric and cometary impacts.

relativity   The theory of space and time developed by the German-Swiss-American physicist Albert Einstein (1879–1955) early in the 20th century. Relativity and quantum theory serve as the two pillars of modern physics.

remote sensing   The sensing of an object or phenomenon, using different portions of the electromagnetic spectrum, without having the sensor in direct contact with the object being studied. In astronomy characteristic electromagnetic radiation signatures often carry distinctive information about an interesting celestial object that is light-years away from the sensor.

resolution   The smallest detail (measurement) that can be distinguished by a sensor system under specific conditions, such as its spatial resolution or spectral resolution.

retrograde motion   Motion in a reverse or backward direction.

rift valley   A depression in a planet's surface due to crustal mass separation.

rille   A deep, narrow depression on the lunar surface that cuts across all other types of topographical features on the Moon.

ring (planetary)   A disk of matter that encircles a planet; such rings usually contain ice and dust particles, ranging in size from microscopic fragments to chunks that are tens of feet in diameter.

robot spacecraft   A semi-automated or fully automated spacecraft capable of executing its primary exploration mission with minimal or no human supervision.

Roche limit   As postulated in the 19th century by the French mathematician Édouard Albert Roche (1820–83), the smallest distance from a planet at which gravitational forces can hold together a satellite or moon that has the same average density as the primary body. If the moon's orbit falls within the Roche limit, it will be torn apart by tidal forces.

rogue star   A wandering star that passes close to a solar system, disrupting the celestial bodies in the system and triggering cosmic catastrophes on life-bearing planets.

satellite   A secondary (smaller) celestial body in orbit around a larger primary body. For example, Earth is a natural satellite of the Sun, while the Moon is a natural satellite of Earth. A human-made spacecraft placed in orbit around Earth is called an artificial satellite—or more commonly just a satellite.

scarp   A cliff produced by erosion or faulting.

science payload   The collection of scientific instruments on a spacecraft.

sensor   The portion of a scientific instrument that detects and/or measures some physical phenomenon.

Seyfert galaxy   A type of spiral galaxy with a very bright galactic nucleus—first observed in 1943 by the American astronomer Carl Keenan Seyfert (1911–60).

shepherd moon   A small inner moon (or pair of moons) that shapes and forms a particular ring around a (ringed) planet; for example, the shepherd moons, Ophelia and Cordelia, tend the Epsilon Ring of Uranus.

shield volcano   A wide, gently sloping volcano formed by the gradual outflow of molten rock.

short-period comet   A comet with an orbital period of less than 200 years.

sidereal   Of or pertaining to stars.

singularity   The hypothetical central point in a black hole at which the curvature of space and time becomes infinite; a theoretical point that has infinite density and zero volume.

SI units   The international system of units (the metric system) that uses the meter (m), kilogram (kg), and second (s) as its basic units of length, mass, and time, respectively.

Small Magellanic Cloud   An irregular galaxy about 9,000 light-years in diameter and 180,000 light-years from Earth. *See also* MAGELLANIC CLOUDS.

soft landing   The act of landing on the surface of a planet without damaging any portion of a spacecraft or its payload, except possibly an expendable landing-gear structure. *Compare with* HARD LANDING.

Sol   The Sun.

sol   A Martian day (about 24 hours, 37 minutes, and 23 seconds in duration). Seven sols equal about 7.2 Earth days.

solar   Of or pertaining to the Sun; caused by the Sun.

solar activity   Any variation in the appearance or energy output of the Sun.

solar constant   The total average amount of the solar energy (in all wavelengths) crossing perpendicular to a unit area at the top of Earth's atmosphere; measured by spacecraft at about 435 Btu/h-ft^2 (1,370 W/m^2) at one astronomical unit from the Sun.

solar flare   A highly concentrated, explosive release of electromagnetic radiation and nuclear particles within the Sun's atmosphere near an active sunspot.

solar mass   The mass of the Sun, about 4.38×10^{30} pounds (1.99×10^{30} kg). It is commonly used as a reference mass in stellar astronomy.

solar nebula   The cloud of dust and gas from which the Sun, the planets, and other minor bodies of this solar system are thought to have formed about 5 billion years ago.

solar system   In general, any star and its gravitationally bound collection of nonluminous objects, such as planets, asteroids, and comets; specifically, humans' home solar system, consisting of the Sun and all the objects bound to it by gravitation—including eight major planets, three dwarf planets (Pluto, Ceres, and 2003 UB313), more than 60 known moons, more than 2,000 asteroids (minor planets), and a very large number of comets. Except for the comets, all the other celestial objects travel around the Sun in the same direction.

solar wind   The variable stream of plasma (i.e., electrons, protons, alpha particles, and other atomic nuclei) that flows continuously outward from the Sun into interplanetary space.

space-based astronomy   The use of astronomical instruments on spacecraft in orbit around Earth and in other locations throughout the solar system to view the universe from above Earth's atmosphere. Major breakthroughs in astronomy, astrophysics, and cosmology have occurred because of the unhampered viewing advantages provided by space platforms.

spacecraft   A platform that can function, move, and operate in outer space or on a planetary surface. Spacecraft can be human-occupied or uncrewed (robot) platforms. They can operate in orbit around Earth or while on an interplanetary trajectory to another celestial body. Some spacecraft travel through space and orbit another planet, while others descend to a planet's surface, making a hard landing (collision impact) or a (survivable) soft landing. Exploration spacecraft are often categorized as either flyby, orbiter, atmospheric probe, lander, or rover spacecraft.

spacecraft clock   The time-keeping component within a spacecraft's command- and data-handling system. It meters the passing time during a mission and regulates nearly all activity within the spacecraft.

spectral classification   The system in which stars are given a designation, consisting of a letter and a number according to their spectral lines, that corresponds roughly to surface temperature. Astronomers classify stars as O (hottest), B, A, F, G, K, and M (coolest). The numbers (zero through nine) represent subdivisions within each major class. The Sun is a G2 star—a little hotter than a G3 star and a little cooler than a G1 star. M stars are numerous but very dim, while O and B stars are very bright but rare.

spectrometer An optical instrument that splits incoming visible light (or other electromagnetic radiation) from a celestial object into a spectrum by diffraction and then measures the relative amplitudes of the different wavelengths. Infrared and ultraviolet spectrometers are often carried on scientific spacecraft.

spectroscopy The study of spectral lines from different atoms and molecules. Astronomers use emission spectroscopy to infer the material composition of the objects that emitted the light and absorption spectroscopy to infer the composition of the intervening medium.

speed of light (symbol: c) The speed at which electromagnetic radiation (including light) moves through a vacuum; a universal constant equal to approximately 186,450 miles (300,000 km) per second.

spin stabilization Directional stability of a spacecraft obtained as a result of spinning the moving body about its axis of symmetry.

spiral galaxy A galaxy with spiral arms, similar to the Milky Way Galaxy or the Andromeda Galaxy.

star A self-luminous ball of very hot gas that liberates energy through thermonuclear fusion reactions within its core. Stars are classified as either normal or abnormal. Normal stars, like the Sun, shine steadily—exhibiting one of a variety of distinctive colors such as red, orange, yellow, blue, and white (in order of increasing surface temperature). There are also several types of abnormal stars, including giant stars; white-, black-, and brown-dwarf stars; and variable stars. Stars experience an evolutionary life cycle from birth in an interstellar cloud of gas to death as a compact white dwarf, neutron star, or black hole.

star cluster A group of stars (numbering from a few to perhaps thousands) that were formed from a common gas cloud and are now bound together by their mutual gravitational attraction.

star probe A conceptual NASA robot scientific spacecraft, capable of approaching within 620,000 miles (1 million km) of the Sun's surface (photosphere) and providing the first in situ measurements of its corona (outer atmosphere).

station keeping The sequence of maneuvers that maintains a space vehicle or spacecraft in a predetermined orbit or on a desired trajectory.

steady-state universe A cosmology model (based on the perfect cosmological principle) suggesting that the universe looks the same to all observers at all times.

stellar evolution The different phases in the lifetime of a star, from its formation out of a interstellar gas and dust, to the time after its nuclear fusion fuel is exhausted.

Sun-like star A yellow, G-spectral classification, main-sequence star with a surface temperature between 5,000 and 6,000 K.

sunspot A relatively dark, sharply defined region on the Sun's visible surface that represents a magnetic area. The sunspot's umbra (darkest region) is about 2,000 K cooler than the effective temperature of the photosphere (some 5,800 K). It is surrounded by a less dark region called the penumbra. Sunspots generally occur in groups of two or more and have diameters ranging from about 2,500 miles (4,000 km) to over 125,000 miles (200,000 km).

Sun-synchronous orbit A very useful polar orbit that allows a satellite's sensor to maintain a fixed relation to the Sun during each local data collection—an important feature for orbiter scientific spacecraft.

supergiant The largest and brightest type of star, with a luminosity of 10,000 to 100,000 times that of the Sun.

supernova The catastrophic explosion of a massive star at the end of its life cycle. As the star collapses and explodes, it experiences a variety of energetic nuclear reactions that lead to the creation of heavier elements, which are then scattered into space. Its brightness increases several million times in a matter of days and outshines all other objects in its galaxy.

Surveyor Project The NASA Moon exploration effort in which five lander spacecraft softly touched down on the lunar surface between 1966 and 1968—the robot precursor to the Apollo Project human expeditions.

tail (cometary) The long, wispy portion of some comets, containing the gas (plasma tail) and dust (dust tail), which stream out of the comet's head (coma) as it approaches the Sun. The plasma tail interacts with the solar wind and points straight back from the Sun, while the dust tail can be curved and fan-shaped.

telecommunications   The transmission of information over great distances using radio waves or other portions of the electromagnetic spectrum.

telemetry   The process of making measurements at one point and transmitting the information via radio waves over some distance to another location for evaluation and use. Telemetered data on a spacecraft's communications downlink often includes scientific data, as well as spacecraft state-of-health data.

telescope   An instrument that collects electromagnetic radiation from a distant object so as to form an image of the object or to permit the radiation signal to be analyzed. Optical (astronomical) telescopes are divided into two general classes: refracting telescopes and reflecting telescopes. Earth-based astronomers also use large radio telescopes, while orbiting observatories use optical, infrared, ultraviolet, X-ray, and gamma-ray telescopes to study the universe.

terminator   The distinctive boundary line separating the illuminated (i.e., sunlit) and dark portions of a nonluminous celestial body like the Moon.

terrestrial planets   In addition to Earth, the planets Mercury, Venus, and Mars—all of which are relatively small, high-density celestial bodies composed of metals and silicates and with shallow or no atmospheres in comparison to the Jovian planets.

Titan   The largest moon of Saturn, discovered in 1655 by the Dutch astronomer Christiaan Huygens (1629–95). It is the only moon in the solar system with a significant atmosphere.

trajectory   The three-dimensional path traced by any object moving because of an externally applied force; the flight path of a space vehicle.

transfer orbit   An elliptical interplanetary trajectory tangent to the orbits of both the departure planet and target planet (or moon).

transit (planetary)   The passage of one celestial body in front of another (larger diameter) celestial body, such as Venus across the face of the Sun.

Trans-Neptunian object (TNO)   Any of the numerous small, icy celestial bodies that lie in the outer fringes of the solar system beyond Neptune. TNOs include plutinos and Kuiper belt objects.

ultraviolet astronomy   The branch of astronomy, conducted primarily from space-based observatories, that uses the ultraviolet portion of the electromagnetic spectrum to study unusual interstellar and intergalactic phenomena.

universal time (UT)   The worldwide civil time standard, equivalent to Greenwich mean time.

uplink   The telemetry signal sent from a ground station to a spacecraft or planetary probe.

Valles Marineris   An extensive canyon system on Mars near the planet's equator, discovered in 1971 by NASA's *Mariner 9* spacecraft.

variable star   A star that does not shine steadily but whose brightness (luminosity) changes over a short period of time.

Venera   The family of robot spacecraft (flybys, orbiters, probes, and landers) from the former Soviet Union that successfully explored Venus, including its inferno-like surface, between 1961 and 1984.

Viking Project   NASA's highly successful Mars exploration effort in the 1970s in which two orbiter and two lander robot spacecraft conducted the first detailed study of the Martian environment and the first (albeit inconclusive) scientific search for life on the Red Planet.

Voyager 1, 2   NASA's twin robot spacecraft that explored the outer regions of the solar system, visiting all the Jovian planets. *Voyager 1* encountered Jupiter (1979) and Saturn (1980) before departing on an interstellar trajectory. *Voyager 2* performed the historic grand tour mission by visiting Jupiter (1979), Saturn (1981), Uranus (1986), and Neptune (1989). Both RTG-powered spacecraft are now involved in the Voyager Interstellar Mission, and each carries a special recording ("Sounds of Earth")—a digital message for any intelligent species that finds them drifting between the stars millennia from now.

white dwarf (star)   A compact star at the end of its life cycle. Once a star of one solar mass or less exhausts its nuclear fuel, it collapses under gravity into a very dense object about the size of Earth.

X-ray A penetrating form of electromagnetic radiation of very short wavelength (approximately 0.01 to 10 nanometers) and high photon energy (approximately 100 electron volts to some 100 kiloelectron volts.)

X-ray astronomy The branch of astronomy, primarily space-based, that uses characteristic X-ray emissions to study very energetic and violent processes throughout universe. X-ray emissions carry information about the temperature, density, age, and other physical conditions of celestial objects that produced them. Objects of interest include supernova remnants, pulsars, active galaxies, and energetic solar flares.

Yohkoh A Japanese solar X-ray observation satellite launched in 1991.

young stellar object Any celestial object in an early stage of star formation, from a protostar to a main-sequence star.

zenith The point on the celestial sphere vertically overhead. *Compare with* NADIR.

Zond A family of robot spacecraft from the former Soviet Union that explored the Moon, Mars, Venus, and interplanetary space in the 1960s.

Further Reading

RECOMMENDED BOOKS

Angelo, Joseph A., Jr. *The Dictionary of Space Technology.* Rev. ed. New York: Facts On File, Inc., 2004.

———. *Encyclopedia of Space Exploration.* New York: Facts On File, Inc., 2000.

———, and Irving W. Ginsberg, eds. *Earth Observations and Global Change Decision Making, 1989: A National Partnership.* Malabar, Fla.: Krieger Publishing, 1990.

Brown, Robert A., ed. *Endeavour Views the Earth.* New York: Cambridge University Press, 1996.

Burrows, William E., and Walter Cronkite. *The Infinite Journey: Eyewitness Accounts of NASA and the Age of Space.* Discovery Books, 2000.

Chaisson, Eric, and Steve McMillian. *Astronomy Today.* 5th ed. Upper Saddle River, N.J.: Pearson Prentice Hall, 2005.

Cole, Michael D. *International Space Station. A Space Mission.* Springfield, N.J.: Enslow Publishers, 1999.

Collins, Michael. *Carrying the Fire.* New York: Cooper Square Publishers, 2001.

Consolmagno, Guy J., et al. *Turn Left at Orion: A Hundred Night Objects to See in a Small Telescope—And How to Find Them.* New York: Cambridge University Press, 2000.

Damon, Thomas D. *Introduction to Space: The Science of Spaceflight.* 3d ed. Malabar, Fla.: Krieger Publishing Co., 2000.

Dickinson, Terence. *The Universe and Beyond.* 3d ed. Willowdater, Ont.: Firefly Books Ltd., 1999.

Heppenheimer, Thomas A. *Countdown: A History of Space Flight.* New York: John Wiley and Sons, 1997.

Kluger, Jeffrey. *Journey beyond Selene: Remarkable Expeditions Past Our Moon and to the Ends of the Solar System.* New York: Simon & Schuster, 1999.

Kraemer, Robert S. *Beyond the Moon: A Golden Age of Planetary Exploration, 1971–1978.* Smithsonian History of Aviation and Spaceflight Series. Washington, D.C.: Smithsonian Institution Press, 2000.

Lewis, John S. *Rain of Iron and Ice: The Very Real Threat of Comet and Asteroid Bombardment.* Reading, Mass.: Addison-Wesley, 1996.

Logsdon, John M. *Together in Orbit: The Origins of International Participation in the Space Station.* NASA History Division, Monographs in Aerospace History 11, Washington, D.C.: Office of Policy and Plans, November 1998.

Matloff, Gregory L. *The Urban Astronomer: A Practical Guide for Observers in Cities and Suburbs.* New York: John Wiley and Sons, 1991.

Neal, Valerie, Cathleen S. Lewis, and Frank H. Winter. *Spaceflight: A Smithsonian Guide.* New York: Macmillan, 1995.

Pebbles, Curtis L. *The Corona Project: America's First Spy Satellites.* Annapolis, Md.: Naval Institute Press, 1997.

Seeds, Michael A. Horizons: *Exploring the Universe.* 6th ed. Pacific Grove, Calif.: Brooks/Cole Publishing, 1999.

Sutton, George Paul. *Rocket Propulsion Elements.* 7th ed. New York: John Wiley & Sons, 2000.

Todd, Deborah, and Joseph A. Angelo, Jr. *A to Z of Scientists in Space and Astronomy.* New York: Facts On File, Inc., 2005.

EXPLORING CYBERSPACE

In recent years, numerous Web sites dealing with astronomy, astrophysics, cosmology, space exploration, and the search for life beyond Earth have appeared on the Internet. Visits to such sites can provide information about the status of ongoing missions, such as NASA's *Cassini* spacecraft as it explores the Saturn system. This book can serve as an important companion, as you explore a new Web site and encounter a person, technology phrase, or physical concept unfamiliar to you and not fully discussed within the particular site. To help enrich the content of this book and to make your astronomy and/or space technology–related travels in cyberspace more enjoyable and productive, the following is a selected list of Web sites that are recommended for your viewing. From these sites you will be able to link to many other astronomy or space-related locations on the Internet. Please note that this is obviously just a partial list of the many astronomy and space-related Web sites now available. Every effort has been made at the time of publication to ensure the accuracy of the information provided. However, due to the dynamic nature of the Internet, URL changes do occur and any inconvenience you might experience is regretted.

Selected Organizational Home Pages

European Space Agency (ESA) is an international organization whose task is to provide for and promote, exclusively for peaceful purposes, cooperation among European states in space research and technology and their applications. URL: http://www.esrin.esa.it. Accessed on April 12, 2005.

National Aeronautics and Space Administration (NASA) is the civilian space agency of the United States government and was created in 1958 by an act of Congress. NASA's overall mission is to plan, direct, and conduct American civilian (including scientific) aeronautical and space activities for peaceful purposes. URL: http://www.nasa.gov. Accessed on April 12, 2005.

National Oceanic and Atmospheric Administration (NOAA) was established in 1970 as an agency within the U.S. Department of Commerce to ensure the safety of the general public from atmospheric phenomena and to provide the public with an understanding of Earth's environment and resources. URL: http://www.noaa.gov. Accessed on April 12, 2005.

National Reconnaissance Office (NRO) is the organization within the Department of Defense that designs, builds, and operates U.S. reconnaissance satellites. URL: http://www.nro.gov. Accessed on April 12, 2005.

United States Air Force (USAF) serves as the primary agent for the space defense needs of the United States. All military satellites are launched from Cape Canaveral Air Force Station, Florida or Vandenberg Air Force Base, California. URL: http://www.af.mil. Accessed on April 14, 2005.

United States Strategic Command (USSTRATCOM) is the strategic forces organization within the Department of Defense, which commands and controls U.S. nuclear forces and military space operations. URL: http://www.stratcom.mil. Accessed on April 14, 2005.

Selected NASA Centers

Ames Research Center (ARC) in Mountain View, California, is NASA's primary center for exobiology, information technology, and aeronautics. URL: http://www.arc.nasa.gov. Accessed on April 12, 2005.

Dryden Flight Research Center (DFRC) in Edwards, California, is NASA's center for atmospheric flight operations and aeronautical flight research. URL: http://www.dfrc.nasa.gov. Accessed on April 12, 2005.

Glenn Research Center (GRC) in Cleveland, Ohio, develops aerospace propulsion, power, and communications technology for NASA. URL: http://www.grc.nasa.gov. Accessed on April 12, 2005.

Goddard Space Flight Center (GSFC) in Greenbelt, Maryland, has a diverse range of responsibilities within NASA, including Earth system science, astrophysics, and operation of the *Hubble Space Telescope* and other Earth-orbiting spacecraft. URL: http://www.nasa.gov/goddard. Accessed on April 14, 2005.

Jet Propulsion Laboratory (JPL) in Pasadena, California, is a government-owned facility operated for NASA by Caltech. JPL manages and operates NASA's deep-space scientific missions, as well as the NASA's Deep Space Network, which communicates with solar system exploration spacecraft. URL: http://www.jpl.nasa.gov. Accessed on April 12, 2005.

Johnson Space Center (JSC) in Houston, Texas, is NASA's primary center for design, development, and testing of spacecraft and associated systems for human space flight, including astronaut selection and training. URL: http://www.jsc.nasa.gov. Accessed on April 12, 2005.

Kennedy Space Center (KSC) in Florida is the NASA center responsible for ground turnaround and support operations, prelaunch checkout, and launch of the space shuttle. This center is also responsible for NASA launch facilities at Vandenberg Air Force Base, California. URL: http://www.ksc.nasa.gov. Accessed on April 12, 2005.

Langley Research Center (LaRC) in Hampton, Virginia, is NASA's center for structures and materials, as well as hypersonic flight research and aircraft safety. URL: http://www.larc.nasa.gov. Accessed on April 15, 2005.

Marshall Space Flight Center (MSFC) in Huntsville, Alabama, serves as NASA's main research center for space propulsion, including contemporary rocket engine development as well as advanced space transportation system concepts. URL: http://www.msfc.nasa.gov. Accessed on April 12, 2005.

Stennis Space Center (SSC) in Mississippi is the main NASA center for large rocket engine testing, including space shuttle engines as well as future generations of space launch vehicles. URL: http://www.ssc.nasa.gov. Accessed on April 14, 2005.

Wallops Flight Facility (WFF) in Wallops Island, Virginia, manages NASA's suborbital sounding rocket program and scientific balloon flights to Earth's upper atmosphere. URL: http://www.wff.nasa.gov. Accessed on April 14, 2005.

White Sands Test Facility (WSTF) in White Sands, New Mexico, supports the space shuttle and space station programs by performing tests on and evaluating potentially hazardous materials, space flight components, and rocket propulsion systems. URL: http://www.wstf.nasa.gov. Accessed on April 12, 2005.

Selected Space Missions

Cassini Mission is an ongoing scientific exploration of the planet Saturn. URL: http://saturn.jpl.nasa.gov. Accessed on April 14, 2005.

Chandra X-ray Observatory (CXO) is a space-based astronomical observatory that is part of NASA's Great Observatories Program. *CXO* observes the universe in the X-ray portion of the electromagnetic spectrum. URL: http://www.chandra.harvard.edu. Accessed on April 14, 2005.

Exploration of Mars is the focus of this Web site, which features the results of numerous contemporary and previous flyby, orbiter, and lander robotic spacecraft. URL: http://mars.jpl.nasa.gov. Accessed on April 14, 2005.

National Space Science Data Center (NSSDC) provides a worldwide compilation of space missions and scientific spacecraft. URL: http://nssdc.gsfc.nasa.gov/planetary. Accessed on April 14, 2005.

Voyager (Deep Space/Interstellar) updates the status of NASA's *Voyager 1* and *2* spacecraft as they travel beyond the solar system. URL: http://voyager.jpl.nasa.gov. Accessed on April 14, 2005.

Other Interesting Astronomy and Space Sites

Arecibo Observatory in the tropical jungle of Puerto Rico is the world's largest radio/radar telescope. URL: http://www.naic.edu. Accessed on April 14, 2005.

Astrogeology (USGS) describes the USGS Astrogeology Research Program, which has a rich history of participation in space exploration efforts and planetary mapping. URL: http://planetarynames.wr.usgs.gov. Accessed on April 14, 2005.

Hubble Space Telescope (**HST**) is an orbiting NASA Great Observatory that is studying the universe primarily in the visible portions of the electromagnetic spectrum. URL: http://hubblesite.org. Accessed on April 14, 2005.

NASA's Deep Space Network (DSN) is a global network of antennas that provide telecommunications support to distant interplanetary spacecraft and probes. URL: http://deepspace.jpl.nasa.gov/dsn. Accessed on April 14, 2005.

NASA's Space Science News provides contemporary information about ongoing space science activities. URL: http://science.nasa.gov. Accessed on April 14, 2005.

National Air and Space Museum (NASM) of the Smithsonian Institution in Washington, D.C., maintains the largest collection of historic aircraft and spacecraft in the world. URL: http://www.nasm.si.edu. Accessed on April 14, 2005.

Planetary Photojournal is a NASA-/JPL- sponsored Web site that provides an extensive collection of images of celestial objects within and beyond the solar system, historic and contemporary spacecraft used in space exploration, and advanced aerospace technologies URL: http://photojournal.jpl.nasa.gov. Accessed on April 14, 2005.

Planetary Society is the nonprofit organization founded in 1980 by Carl Sagan and other scientists that encourages all spacefaring nations to explore other worlds. URL: http://planetary.org. Accessed on April 14, 2005.

Search for Extraterrestrial Intelligence (SETI) Projects at UC Berkeley is a Web site that involves contemporary activities in the search for extraterrestrial intelligence (SETI), especially a radio SETI project that lets anyone with a computer and an Internet connection participate. URL: http://www.setiathome.ssl.berkeley.edu. Accessed on April 14, 2005.

Solar System Exploration is a NASA-sponsored and -maintained Web site that presents the last events, discoveries and missions involving the exploration of the solar system. URL: http://solarsystem.nasa.gov. Accessed on April 14, 2005.

Space Flight History is a gateway Web site sponsored and maintained by the NASA Johnson Space Center. It provides access to a wide variety of interesting data and historic reports dealing with (primarily U.S.) human space flight. URL: http://www11.jsc.nasa.gov/history. Accessed on April 14, 2005.

Space Flight Information (**NASA**) is a NASA-maintained and -sponsored gateway Web site that provides the latest information about human spaceflight activities, including the *International Space Station* and the space shuttle. URL: http://spaceflight.nasa.gov Accessed on April 14, 2005.

Index